Christopher Albler

Metals in Carbohydrate Synthesis

Christopher Albler

Metals in Carbohydrate Synthesis

Indium and Titanium mediated Chain Elongations of Chiral Pool Building Blocks

Südwestdeutscher Verlag für Hochschulschriften

Impressum / Imprint

Bibliografische Information der Deutschen Nationalbibliothek: Die Deutsche Nationalbibliothek verzeichnet diese Publikation in der Deutschen Nationalbibliografie; detaillierte bibliografische Daten sind im Internet über http://dnb.d-nb.de abrufbar.

Alle in diesem Buch genannten Marken und Produktnamen unterliegen warenzeichen-, marken- oder patentrechtlichem Schutz bzw. sind Warenzeichen oder eingetragene Warenzeichen der jeweiligen Inhaber. Die Wiedergabe von Marken, Produktnamen, Gebrauchsnamen, Handelsnamen, Warenbezeichnungen u.s.w. in diesem Werk berechtigt auch ohne besondere Kennzeichnung nicht zu der Annahme, dass solche Namen im Sinne der Warenzeichen- und Markenschutzgesetzgebung als frei zu betrachten wären und daher von jedermann benutzt werden dürften.

Bibliographic information published by the Deutsche Nationalbibliothek: The Deutsche Nationalbibliothek lists this publication in the Deutsche Nationalbibliografie; detailed bibliographic data are available in the Internet at http://dnb.d-nb.de.

Any brand names and product names mentioned in this book are subject to trademark, brand or patent protection and are trademarks or registered trademarks of their respective holders. The use of brand names, product names, common names, trade names, product descriptions etc. even without a particular marking in this work is in no way to be construed to mean that such names may be regarded as unrestricted in respect of trademark and brand protection legislation and could thus be used by anyone.

Coverbild / Cover image: www.ingimage.com

Verlag / Publisher:
Südwestdeutscher Verlag für Hochschulschriften
ist ein Imprint der / is a trademark of
OmniScriptum GmbH & Co. KG
Heinrich-Böcking-Str. 6-8, 66121 Saarbrücken, Deutschland / Germany
Email: info@svh-verlag.de

Herstellung: siehe letzte Seite /
Printed at: see last page
ISBN: 978-3-8381-5089-5

Zugl. / Approved by: Wien, Universität Wien, Diss., 2014

Copyright © 2015 OmniScriptum GmbH & Co. KG
Alle Rechte vorbehalten. / All rights reserved. Saarbrücken 2015

Table of Contents

List of abbreviations 5

1 Introduction 9
   1.1 Barbier-type indium reagents in organic synthesis 10
      1.1.2 Indium mediated carbonyl allylation 12
      1.1.2 Propargylation and allenylation 15
      1.1.3 Indium mediated alkylation 18
   1.2 Titanium mediated aldol-type reactions in organic synthesis 19
      1.2.1 The Mukaiyama aldol reaction 19
      1.2.2 The Evans aldol reaction 25
      1.2.3 Garner's aldehyde 33
      1.2.4 Aldol additions in carbohydrate synthesis 34

2 Results and discussion 40
   2.1 Aim and background of the projects 40
   2.2 Motivation for the projects 43
      2.2.1 Motivation for the synthesis of higher amino sugars 43
      2.2.2 Motivation for the synthesis of fluorinated amino sugars 45
   2.3 Synthesis of higher amino sugars 49
      2.3.1 Indium mediated allylation of unprotected carbohydrates 49
      2.3.2 Epoxidation of unsaturated aldehydes 51
      2.3.3 Nucleophilic azide opening of epoxides 53
      2.3.4 Deprotection protocol 55
   2.4 Synthesis of fluorinated amino sugars 63
      2.4.1 Epoxidation/fluoride opening approach 63
      2.4.2 Allylation/ozonolysis/α-fluorination approach 66
      2.4.3 Stereoselective aldol addition approach 69
      2.4.4 Further elaboration on the aldol addition approach 81

3 Experimental part 94
   3.1 General methods 94
   3.2 General procedures 96
      Method A: Acidic deacetylation, ozonolysis. Synthesis of 6a-c 96
      Method B: Azide reduction. Synthesis of 7a-7c 96
      Method C: Zemplén saponification. Synthesis of 8a-c, 28-31 97
      Method D: Aldol addition. Synthesis of 15-19, 32-33, 42 97
      Method E: Oxazolidinone, acetonide cleavage. Synthesis of 20-23 98
      Method F: DIBAL reduction, Boc cleavage. Synthesis of 24-27 98
      Method G: Reductive auxiliary cleavage. Synthesis of 34-35 99
      Method H: Pummerer rearrangement, acidic deprotection. Synthesis of 40-41 99
      Procedure for the preparation of the Roush reagent 100
      Procedure for the preparation of flouroacetyl chloride 101
      Procedure for epimerization of L-threonine to D-*allo*-threonine 101
   3.4 Experimental procedures and data for key intermediates and final products[1,2] ... 102

2-azido-2-deoxy-D-glycero-D-ido-heptose (6a) 102
2-azido-2-deoxy-D-threo-L-galacto-octose (6b) 102
2-azido-2-deoxy-D-erythro-L-galacto-octose (6c) 103
2-acetamido-1,3,4,6,7-penta-O-acetyl-2-deoxy-D-glycero-D-ido-heptose (7a) 104
2-acetamido-1,3,4,6,7,8-hexa-O-acetyl-2-deoxy-D-threo-L-galacto-octose (7b) 106
2-acetamido-1,3,4,6,7,8-hexa-O-acetyl-2-deoxy-D-erythro-L-galacto-octose (7c) 107
2-acetamido-2-deoxy-D-glycero-D-ido-heptose (8a) 108
2-acetamido-2-deoxy-D-threo-L-galacto-octose (8b) 109
2-acetamido-2-deoxy-D-erythro-L-galacto-octose (8c) 110
(R,E)-methyl 4-(dibenzylamino)-5-((4-methoxybenzyl)oxy)pent-2-enoate (9a) 111
(R,E)-methyl 5-((tert-butyldiphenylsilyl)oxy)-4-(dibenzylamino)pent-2-enoate (9b) 112
(R,E)-methyl 4-(dibenzylamino)-5-(pivaloyloxy)pent-2-enoate (9c) 112
((2S,3S)-3-((S)-1-(dibenzylamino)-2-((4-methoxybenzyl)oxy)ethyl)oxiran-2-yl)methanol (10) 113
(2R,3S)-methyl 3-((S)-2-((tert-butyldiphenylsilyl)oxy)-1-(dibenzylamino)ethyl)oxirane-2-carboxylate (11)
.... 114
(S)-tert-butyl 4-((R)-1-(benzyloxy)-3-oxopropyl)-2,2-dimethyloxazolidine-3-carboxylate (12) 115
(R)-tert-butyl 4-((1R,2S,E)-1-(benzyloxy)-5-ethoxy-2-fluoro-5-oxopent-3-en-1-yl)-2,2-
dimethyloxazolidine-3-carboxylate (13) 116
(4S,5R)-3-(2-fluoroacetyl)-4-methyl-5-phenyloxazolidin-2-one (14) 117
(S)-tert-butyl-4-((1S,2R)-2-fluoro-1-hydroxy-3-((4S,5R)-4-methyl-2-oxo-5-phenyloxazolidin-3-yl)-3-
oxopropyl)-2,2-dimethyloxazolidine-3-carboxylate (15) 118
(S)-tert-butyl-4-((1R,2S)-2-fluoro-1-hydroxy-3-((4S,5R)-4-methyl-2-oxo-5-phenyloxazolidin-3-yl)-3-
oxopropyl)-2,2-dimethyloxazolidine-3-carboxylate (16) 119
4R,5R)-tert-butyl-4-((1R,2S)-2-fluoro-1-hydroxy-3-((4S,5R)-4-methyl-2-oxo-5-phenyloxazolidin-3-yl)-3-
oxopropyl)-2,2,5-trimethyloxazolidine-3-carboxylate (17) 120
(4S,5S)-tert-butyl 4-((1R,2S)-2-fluoro-1-hydroxy-3-((4S,5R)-4-methyl-2-oxo-5-phenyloxazolidin-3-yl)-3-
oxopropyl)-2,2,5-trimethyloxazolidine-3-carboxylate (18) 121
(4S,5R)-tert-butyl-4-((1R,2S)-2-fluoro-1-hydroxy-3-((4S,5R)-4-methyl-2-oxo-5-phenyloxazolidin-3-yl)-3-
oxopropyl)-2,2,5-trimethyloxazolidine-3-carboxylate (19) 122
(2S,3R,4S)-methyl-4-((tert-butoxycarbonyl)amino)-2-fluoro-3,5-dihydroxypentanoate (20) 122
(2S,3R,4R)-methyl-4-((tert-butoxycarbonyl)amino)-2-fluoro-3,5-dihydroxypentanoate (21) 123
(2S,3R,4S,5R)-methyl-4-((tert-butoxycarbonyl)amino)-2-fluoro-3,5-dihydroxyhexanoate (22) 124
(2S,3R,4R,5S)-methyl-4-((tert-butoxycarbonyl)amino)-2-fluoro-3,5-dihydroxyhexanoate (23) 125
4-acetamido-1,3-di-O-acetyl-2,4-dideoxy-2-fluoro-D-xylose (24) 125
4-acetamido-1-O-acetyl-2,4-dideoxy-2-fluoro-D-arabinose (25) 126
4-acetamido-1,3-di-O-acetyl-2,4,6-trideoxy-2-fluoro-D-idose (26) 127
4-acetamido-1-O-acetyl-2,4,6-trideoxy-2-fluoro-L-galactose (27) 128
4-acetamido-2,4-dideoxy-2-fluoro-D-xylose (28) 129
4-acetamido-2,4-dideoxy-2-fluoro-D-arabinose (29) 130
4-acetamido-2,4,6-trideoxy-2-fluoro-D-idose (30) 131
4-acetamido-2,4,6-trideoxy-2-fluoro-D-galactose (31) 132
(S)-tert-butyl 4-((1R,2S)-2-fluoro-1-hydroxy-3-((4S,5R)-4-methyl-2-oxo-5-phenyloxazolidin-3-yl)-3-
oxopropyl)-2,2-dimethylthiazolidine-3-carboxylate (32) 133
(R)-tert-butyl 4-((1R,2S)-2-fluoro-1-hydroxy-3-((4S,5R)-4-methyl-2-oxo-5-phenyloxazolidin-3-yl)-3-
oxopropyl)-2,2-dimethylthiazolidine-3-carboxylate (33) 134
(S)-tert-butyl 4-((1R,2R)-2-fluoro-1,3-dihydroxypropyl)-2,2-dimethylthiazolidine-3-carboxylate (34) .. 135
(R)-tert-butyl 4-((1R,2R)-2-fluoro-1,3-dihydroxypropyl)-2,2-dimethylthiazolidine-3-carboxylate (35).. 135
(2R,3R,4R)-4-((2,4-dinitrophenyl)amino)-5-((2,4-dinitrophenyl)thio)-2-fluoropentane-1,3-diol (36) 136
2-acetamido-2,4-dideoxy-4-fluoro-D-lyxose (40) 137
2-acetamido-2,4-dideoxy-4-fluoro-D-xylose (41) 138

(2R,3R,4R,5R,6S)-6-fluoro-5-hydroxy-7-((4S,5R)-4-methyl-2-oxo-5-phenyloxazolidin-3-yl)-7-oxoheptane-1,2,3,4-tetrayl tetraacetate (42) 139

4 NMR Spectra 141

2-azido-2-deoxy-D-glycero-D-ido-heptose (6a) 142

2-azido-2-deoxy-D-threo-L-galacto-octose (6b) 143

2-azido-2-deoxy-D-erythro-L-galacto-octose (6c) 144

2-acetamido-1,3,4,6,7-penta-O-acetyl-2-deoxy-D-glycero-D-ido-heptose (7a) 145

2-acetamido-1,3,4,6,7,8-hexa-O-acetyl-2-deoxy-D-threo-L-galacto-octose (7b) 147

2-acetamido-1,3,4,6,7,8-hexa-O-acetyl-2-deoxy-D-erythro-L-galacto-octose (7c) 148

2-acetamido-2-deoxy-D-glycero-D-ido-heptose (8a) 149

2-acetamido-2-deoxy-D-threo-L-galacto-octose (8b) 150

2-acetamido-2-deoxy-D-erythro-L-galacto-octose (8c) 151

(R,E)-methyl 4-(dibenzylamino)-5-((4-methoxybenzyl)oxy)pent-2-enoate (9a) 152

(R,E)-methyl 4-(dibenzylamino)-5-(pivaloyloxy)pent-2-enoate (9c) 153

((2S,3S)-3-((S)-1-(dibenzylamino)-2-((4-methoxybenzyl)oxy)ethyl)oxiran-2-yl)methanol (10) 154

(2R,3S)-methyl 3-((S)-2-((tert-butyldiphenylsilyl)oxy)-1-(dibenzylamino)ethyl)oxirane-2-carboxylate (11) 155

(R)-tert-butyl 4-((1R,2S,E)-1-(benzyloxy)-5-ethoxy-2-fluoro-5-oxopent-3-en-1-yl)-2,2-dimethyloxazolidine-3-carboxylate (13) 156

(4S,5R)-3-(2-fluoroacetyl)-4-methyl-5-phenyloxazolidin-2-one (14) 157

(S)-tert-butyl-4-((1S,2R)-2-fluoro-1-hydroxy-3-((4S,5R)-4-methyl-2-oxo-5-phenyloxazolidin-3-yl)-3-oxopropyl)-2,2-dimethyloxazolidine-3-carboxylate (15) 158

(S)-tert-butyl-4-((1R,2S)-2-fluoro-1-hydroxy-3-((4S,5R)-4-methyl-2-oxo-5-phenyloxazolidin-3-yl)-3-oxopropyl)-2,2-dimethyloxazolidine-3-carboxylate (16) 159

4R,5R)-tert-butyl-4-((1R,2S)-2-fluoro-1-hydroxy-3-((4S,5R)-4-methyl-2-oxo-5-phenyloxazolidin-3-yl)-3-oxopropyl)-2,2,5-trimethyloxazolidine-3-carboxylate (17) 160

(4S,5S)-tert-butyl 4-((1R,2S)-2-fluoro-1-hydroxy-3-((4S,5R)-4-methyl-2-oxo-5-phenyloxazolidin-3-yl)-3-oxopropyl)-2,2,5-trimethyloxazolidine-3-carboxylate (18) 161

(4S,5R)-tert-butyl-4-((1R,2S)-2-fluoro-1-hydroxy-3-((4S,5R)-4-methyl-2-oxo-5-phenyloxazolidin-3-yl)-3-oxopropyl)-2,2,5-trimethyloxazolidine-3-carboxylate (19) 162

(2S,3R,4S)-methyl-4-((tert-butoxycarbonyl)amino)-2-fluoro-3,5-dihydroxypentanoate (20) 163

(2S,3R,4R)-methyl-4-((tert-butoxycarbonyl)amino)-2-fluoro-3,5-dihydroxypentanoate (21) 164

(2S,3R,4S,5R)-methyl-4-((tert-butoxycarbonyl)amino)-2-fluoro-3,5-dihydroxyhexanoate (22) 165

(2S,3R,4R,5S)-methyl-4-((tert-butoxycarbonyl)amino)-2-fluoro-3,5-dihydroxyhexanoate (23) 166

4-acetamido-1,3-di-O-acetyl-2,4-dideoxy-2-fluoro-D-xylose (24) 167

4-acetamido-1-O-acetyl-2,4-dideoxy-2-fluoro-D-arabinose (25) 168

4-acetamido-1,3-di-O-acetyl-2,4,6-trideoxy-2-fluoro-D-idose (26) 169

4-acetamido-1-O-acetyl-2,4,6-trideoxy-2-fluoro-L-galactose (27) 170

4-acetamido-2,4-dideoxy-2-fluoro-D-xylose (28) 171

4-acetamido-2,4-dideoxy-2-fluoro-D-arabinose (29) 172

4-acetamido-2,4,6-trideoxy-2-fluoro-D-idose (30) 173

4-acetamido-2,4,6-trideoxy-2-fluoro-D-galactose (31) 174

(S)-tert-butyl 4-((1R,2S)-2-fluoro-1-hydroxy-3-((4S,5R)-4-methyl-2-oxo-5-phenyloxazolidin-3-yl)-3-oxopropyl)-2,2-dimethylthiazolidine-3-carboxylate (32) 175

(R)-tert-butyl 4-((1R,2S)-2-fluoro-1-hydroxy-3-((4S,5R)-4-methyl-2-oxo-5-phenyloxazolidin-3-yl)-3-oxopropyl)-2,2-dimethylthiazolidine-3-carboxylate (33) 176

(S)-tert-butyl 4-((1R,2R)-2-fluoro-1,3-dihydroxypropyl)-2,2-dimethylthiazolidine-3-carboxylate (34) 177

(R)-tert-butyl 4-((1R,2R)-2-fluoro-1,3-dihydroxypropyl)-2,2-dimethylthiazolidine-3-carboxylate (35) 178

(2R,3R,4R)-4-((2,4-dinitrophenyl)amino)-5-((2,4-dinitrophenyl)thio)-2-fluoropentane-1,3-diol (36) 179

2-acetamido-2,4-dideoxy-4-fluoro-D-lyxose (40) 180

2-acetamido-2,4-dideoxy-4-fluoro-D-xylose (41) 181

(2R,3R,4R,5R,6S)-6-fluoro-5-hydroxy-7-((4S,5R)-4-methyl-2-oxo-5-phenyloxazolidin-3-yl)-7-oxoheptane-1,2,3,4-tetrayl tetraacetate (42) 182

5 References..182

List of abbreviations

Ac	acetyl
All	allyl
Ar	aryl
Aux	auxiliary
B	base
Bn	benzyl
Boc	*tert*-butyloxycarbonyl
BuLi	*n*-butyl lithium
*t*BuOOH	*tert*-butyl hydroperoxide
Cbz	benzyloxycarbonyl
*m*CPBA	*meta*-chloroperoxybenzoic acid
DAST	diethalaminosulfur trifluoride
DBAD	di-*tert*-butyl azodicarboxylate
DBU	diazabicycloundecene
DCC	dicyclohexylcarbodiimide
DCM	dichloromethane
DHAP	dihydroxyacetone phosphate
DIBAL	diisobutylaluminium hydride
DIPA	diisopropylamine
DIPEA	diisopropylethylamine
DIPT	diisopropyl tartrate
DMAP	dimethylaminopyridine
DMF	dimethylformamide
DMP	dimethoxypropane
DMS	dimethylsulfide
DMSO	dimethylsulfoxide
DTT	dithiothreitol
EA	ethyl acetate

equiv	equivalent
ESI	electrospray ionization
EtOH	ethanol
HE	hexanes
HPLC	high pressure liquid chromatography
HRMS	high resolution mass spectroscopy
HYTRA	2-hydroxy-1,2,2-triphenylethyl acetate
KDN	keto-deoxy-nonulosonic acid
KDO	keto-deoxy-octulosonic acid
L	ligand
LA	Lewis acid
LAH	lithium aluminium hydride
LDA	lithium diisopropylamide
LPS	lipopolysaccharide
MAR	Mukaiyama aldol reaction
MeCN	acetonitrile
MeOH	methanol
MTBE	*tert*-butyl methyl ether
NaHMDS	sodium hexamethyldisilazane
NANA	N-acetyl neuraminic acid
NBS	N-bromosuccinimide
NFSI	N-fluorobenzenesulfonimide
NMO	N-methylmorpholine N-oxide
NMR	nuclear magnetic resonance
NOE	nuclear Overhauser effect
PG	protecting group
Phth	phthalimide
Piv	pivaloyl
PMB	*para*-methoxybenzyl
*i*PrOH	*iso*-propanol

pyr	pyridine
RAMA	rabbit muscle aldolase
RNA	ribonucleic acid
Sia	1,2-dimethylpropyl
SEE	silylenolether
TBAF	tetrabutylammonium fluoride
TBAI	tetrabutylammonium iodide
TBS	*tert*-butyldimethylsilyl
TBDPS	*tert*-butyldiphenylsilyl
TEA	triethylamine
TIPS	triisopropylsilyl
Tf	trifluoromethanesulfon
TFA	trifluoroacetic acid
TFAA	trifluoroacetic anhydride
THF	tetrahydrofurane
TLC	thin-layer chromatography
TMEDA	tetramethylethylenediamine
TMS	trimethylsilyl
TPP	tetraphenyl porphyrin
Ts	*para*-toluenesulfone
TS	transition state
VMAR	vinylogous Mukaiyama aldol reaction

1 Introduction

Carbohydrates in addition to proteins comprise some of the most important molecules of life and due to their huge structural diversity the synthesis of these compounds has always been a challenge for chemists. Especially the preparation of rare and unnatural monosaccharides starting from readily available chiral pool substances such as pentoses, hexoses and amino acids remains an important topic not only in the field of carbohydrate chemistry but also in biochemistry and molecular biology. In order to perform systematic studies for a better understanding of the precise mode of action and biological function of rare carbohydrates such as amino-functionalized and carbon chain elongated sugars, high yielding synthetic routes for their preparation are essential. Additionally, unnatural derivatives are made accessible which are of interest concerning the development of antimicrobial agents and carbohydrate vaccines for the treatment of medical conditions such as viral infections, diabetes or cancer. Owing to their unique properties, metal-containing reagents are used extensively in all fields of the life sciences. Among other applications, they are used as catalysts, redox active reagents, Lewis acids and for the generation of Barbier-type reagents in organic chemistry. In this thesis, the latter two aspects of titanium and indium were harnessed to devise synthetic routes on the one hand towards amino-functionalized heptoses and octoses and on the other hand towards amino-fluoro pentoses and hexoses. Our goal was to develop short and simple synthetic routes which should be flexible in terms of stereochemical variations, thus allowing to synthesize a broad range of substances for biological testing. The first project represents an extension of the indium mediated allylation protocol for unprotected carbohydrates. In this respect, not only carbon chain elongation was performed, but additional functionalization with nitrogen was achieved *via* an epoxidation, azide opening strategy[1]. The second project described herein makes use of titanium mediated aldol additions of amino acid and fluoroacetyl-oxazolidinone derived building blocks for the preparation of fluorinated amino sugars[2] which represent valuable probes for biochemical investigations.

1.1 Barbier-type indium reagents in organic synthesis

The application of the main group metal indium in organometallic reactions has a long history which can be traced back to 1974 when Rieke *et al* performed Reformatsky reactions of ethyl bromoacetate with indium which was prepared by reduction of $InCl_3$ with potassium[3]. Fourteen years later in 1988 Araki and coworkers found that indium powder readily reacted with allylbromide in DMF or THF to generate organometallic species which furnished homoallylic alcohols when treated with carbonyl compounds[4]. Since then indium has drawn a considerable amount of attention among the synthetic community especially with the discovery that many of the reactions could be carried out in aqueous media[5] and that the reagents are tolerant to a number of functional groups. A sesquihalide structure (Scheme 1) was proposed for the indium intermediate in organic solvents whereas in water a surface reaction[6] was anticipated at first since the organo-indium species where thought to easily hydrolyze.

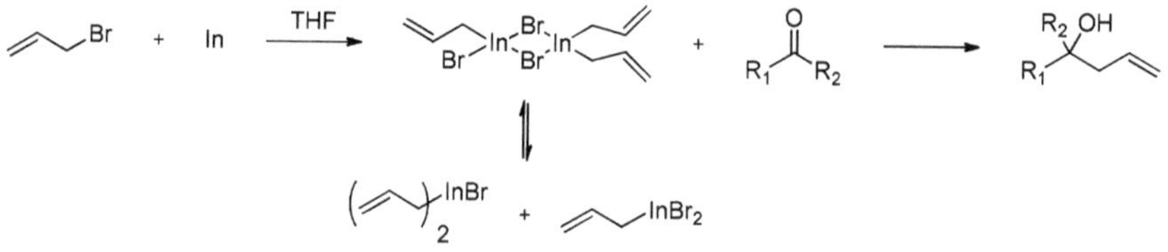

Scheme 1. In mediated allylation in organic media.

In 1993 Whitesides *et al*[7] reported on the indium mediated allylation of unprotected carbohydrates in aqueous media and it was assumed that discrete organometallic species must be involved since separately prepared organoindium reagents gave the same experimental results compared to heterogeneous conditions. Further proof for this hypothesis was brought forward in 1999 when Chan and coworkers performed NMR based investigations of allyl bromide and indium in D_2O^8. A new set of allylic protons was found to emerge rapidly at 1.7 ppm which slowly declined over night leaving allyl alcohol. This experiment also excluded the existence of sesquihalide species which would necessitate two sets of allylic signals. Additionally, transmetallation experiments with diallyl mercury and In^0, InI respectively $InBr_3$ were conducted in order to determine the oxidation state of indium (Scheme 2). Since only In^0 and InI generated the characteristic signal at 1.7 ppm it was concluded that allylindium(I) must be the reactive species in water.

Scheme 2. In mediated allylation in aqueous media.

Following the huge success of the allylation protocol, numerous other transformations with indium have been realized since then[9]. The appealing aspects of indium chemistry can be summarized as follows: Owing to the mildness of organoindium reagents side-reactions such as eliminations and tedious protecting group chemistry can be avoided due to the toleration of many functional groups such hydroxyl moieties, a feature which is especially valuable in carbohydrate chemistry. Additionally the reactions can be carried out in environmentally benign solvents such as alcohols and water and the waste produced is nontoxic. The only disadvantage which might be considered is the cost of indium metal which is mostly required in super-stoichiometric amounts. However, catalytic approaches which for example use manganese as a reducing agent[10] have been developed to overcome this problem.

1.1.2 Indium mediated carbonyl allylation

As mentioned above, Barbier-type reactions can be carried out using indium metal and allyl bromide or iodide in aqueous or organic media. No activation of the metal is required and halogen-metal insertion produces sesquihalide In(III) respectively allylindium(I) species in the two different media. The stoichiometry usually applied is allyl halide/indium/carbonyl = 3/2/1. Additionally ultrasonication[11] and the addition of protic acids[12] were found to lead to better performances. Contributions concerning the investigation of regio- and stereoselectivity were made by the groups of Loh[13] and Paquette[14]. In general the behavior observed was in agreement with existing stereochemical models. On the one hand γ-substituted allyl-halides furnish γ-adducts with *anti* or *syn* diastereoselectivity (Scheme 3), depending on the geometry of the double bond. This behavior reflects the involvement of a six-membered chair-like transition-state, the Zimmerman-Traxler transition state[15] (Scheme 4).

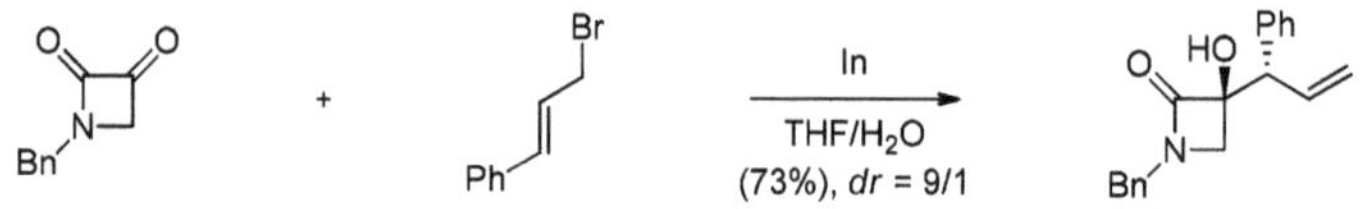

Scheme 3. In mediated allylation of an α-keto-γ-lactam[14a]; γ-regio- and *anti*-diastereoselectivity.

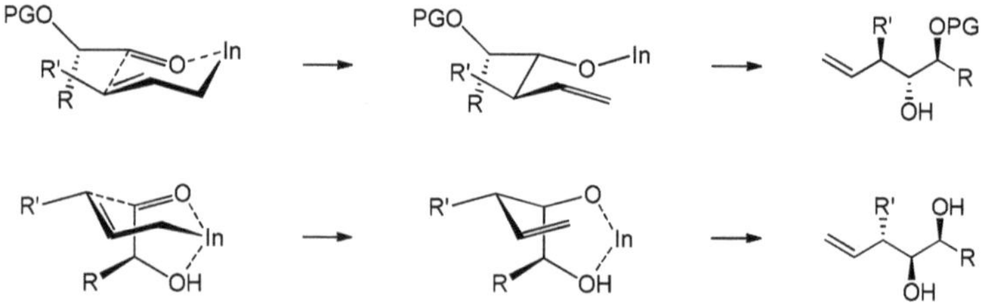

Scheme 4. Chelating and non-chelating Zimmerman-Traxler transition states in the indium mediated allylation.

However, the overall selectivity encountered can be moderate[16] owing to the isomerisation of the crotyl-indium species in the course of the reaction (Scheme

5) which leads to the formation of the respective α-adducts under thermodynamic conditions[17].

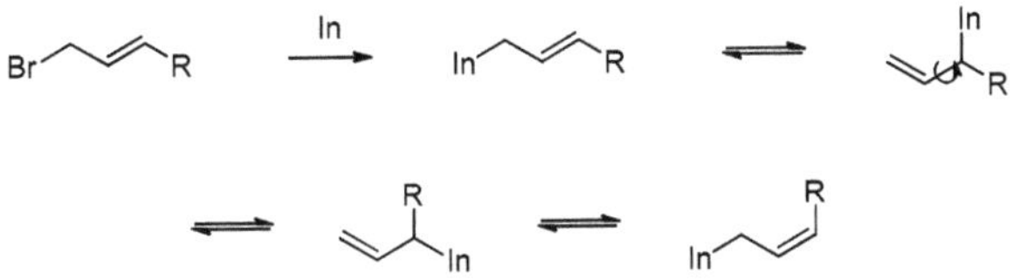

Scheme 5. Isomerisation of crotyl-indium species.

On the other hand α- and β-chiral aldehydes, depending on the nature of the substituent at the chiral center (chelating vs. non-chelating), produce either *syn* or *anti* homoallyl alcohols [18] following the Chelat-Cram and Felkin-Ahn stereochemical models (Scheme 4, Scheme 6).

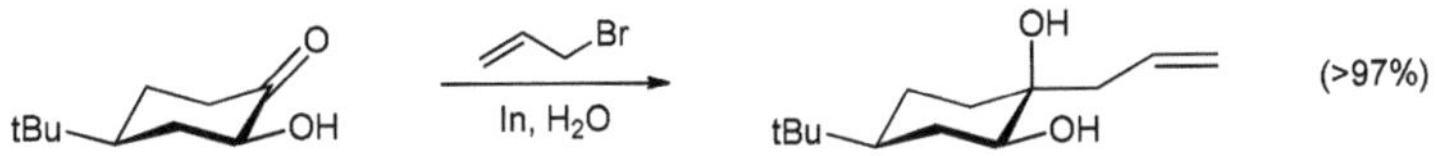

Scheme 6. In mediated allylation of an α-hydroxy ketone[14b]; *syn*-diastereoselectivity.

Allyl-indium reagents are in general hard nucleophiles and therefore lead to 1,2-additions with Michael acceptors[19]. However, when substrates with two electron withdrawing groups[20] or additives such as TMSCl[21] are used, 1,4-additions are preferred. Since indium is able to coordinate to oxygen and nitrogen even in aqueous media, the addition of chiral ligands in the allylation protocol enabled the preparation of enantiomerically enriched products[22] (Scheme 7).

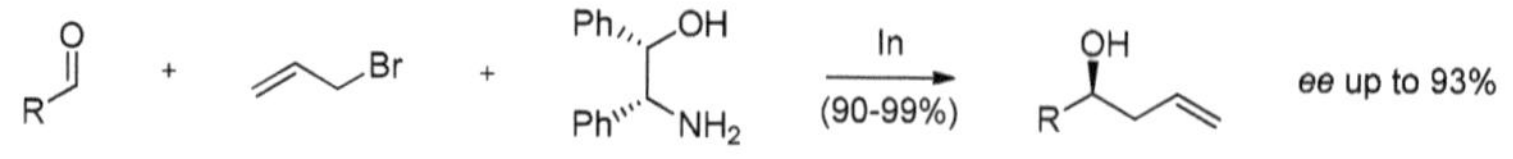

Scheme 7. In promoted allylation with chiral ligands.

Furthermore, allyl-indium reagents have been reacted with a range of electrophiles such as imines [23], nitriles [24] and epoxides [25]. Concerning the application of indium mediated allylations in natural product synthesis the preparation of nonulosonic acids from commercially available hexoses by Chan and Li as well as Whitesides *et al* has to be emphasized (Scheme 8).

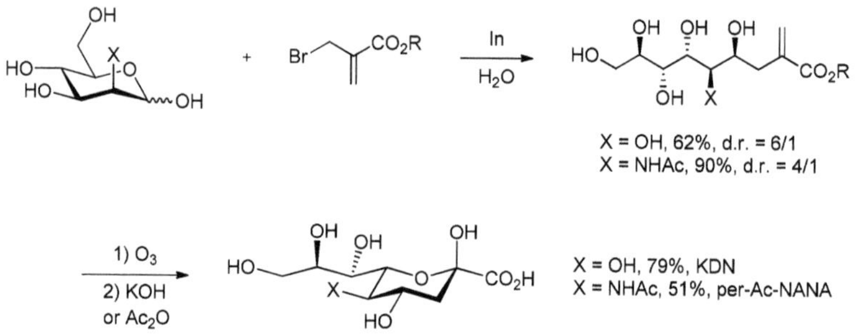

Scheme 8. Synthesis of KDN and NANA *via* indium mediated allylation of hexoses with bromomethyl acrylates.

The synthesis of KDN was improved later on by Fressner *et al* who performed the allylation step in acidic media, improving the overall yield to 75% of a single diastereomer[26]. Interestingly, also the free bromomethyl acrylic acid can be used in the allylation step[27]. Additionally, phosphono-acrylates have been used for the preparation of phosphonate analogues of KDN and NANA, which exert moderate sialidase inhibition.[28] For the preparation of KDO a similar procedure was adopted; starting from diisopropylidene protected arabinose, a *dr* of 2/1 in favor of the *anti* diastereomer[29] was obtained. To further demonstrate the usefulness of indium in carbohydrate synthesis, the preparation of pseudaminic acid from 2,4-diacetamido-2,4,6-trideoxy-L-altrose by Lee and coworkers should be mentioned (Scheme 9)[30].

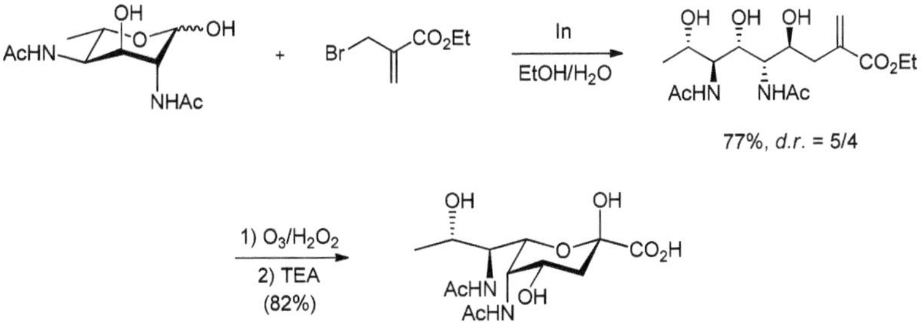

Scheme 9. Synthesis of pseudaminic acid by Lee *et al.*

Although the diastereoselectivity of the allylation step was low and different Lewis acids and chiral ligands failed to improve the selectivity, the required *erythro* isomer was obtained in a slight excess and subsequent ozonolysis and saponification furnished the desired pseudaminic acid, which is associated with the pathogenicity of bacteria[31]. This approach represented the first high yielding synthesis of bacterial sialic acids, although the simple and selective preparation of the required 2,4-diacetamido-2,4,6-trideoxy-hexose precursors still remains a challenge[32].

1.1.2 Propargylation and allenylation

The application of propargyl halides and indium in Barbier-type reactions was first described by Whitesides *et al* in 1993 for the allenylation of unprotected carbohydrates[33].

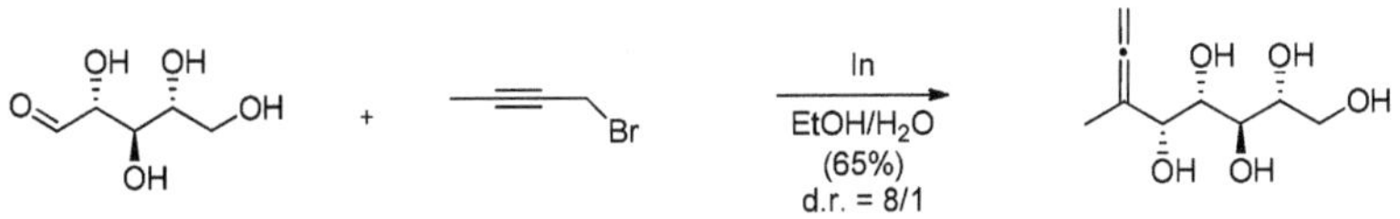

Scheme 10. Allenylation of D-ribose by Whitesides *et al.*

The nature of the organoindium species involved was investigated by Chan *et al*[34]. In analogy to the allyl system, the propargyl/allenyl indium system was found to have a strong dependence on the nature of the solvent used. When propargyl bromide was treated with indium in water, allenylindium(I) was formed, whereas THF promoted the formation of allenylindium(I)/(III) mixtures by NMR analysis. However, when internal alkynes were used, the equilibrium was shifted towards propargylindium species owing to steric repulsion. Thus, substituted propargyl bromides led to the predominant formation of allenyl alcohols and unsubstituted ones furnished the corresponding homopropargylic alcohols[35] (Scheme 11).

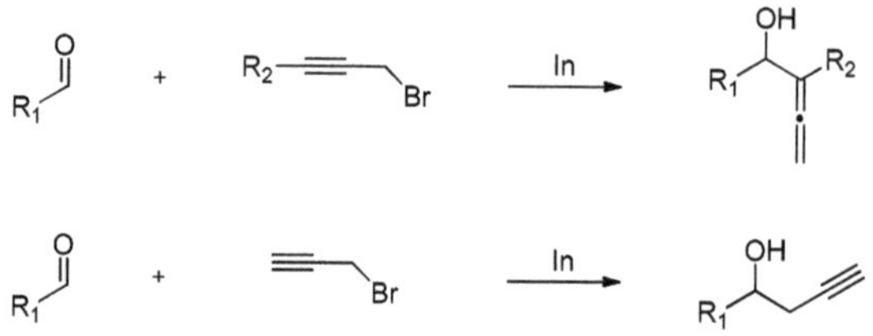

Scheme 11. Indium mediated allenylation/propargylation.

In 2003, Lin and Loh reported on a tunable reaction by using trialkylsilyl propargyl bromides[36]. When TMS propargyl bromide and indium were reacted with aldehydes in the presence of $InBr_3$ in THF, exclusive formation of homopropargylic alcohols was observed. On the other hand, TBDPS propargyl bromide furnished allenyl alcohols (Scheme 12).

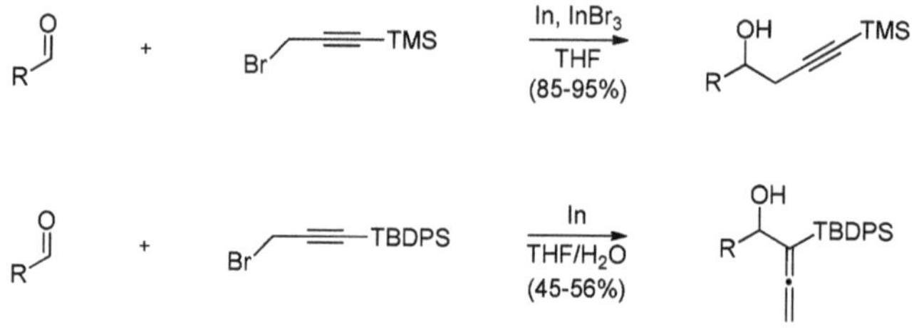

Scheme 12. Regioselective propargylation/allenylation with silyl propargyl bromides.

It was suggested that a chelation between silicon and bromide shifted the equilibrium towards allenylindium species, which was not possible in the case of the more bulky TBDPS moiety. In 2011 Schmid *et al* reported on the allenylation/propargylation of isopropylidene glyceraldehyde with protected 4-bromo-2-butyn-1-ols[37]. In this case, considerable amounts of homopropargylic alcohols were obtained, which was explained by a chelation between indium and oxygen. Therefore, the sterically unfavorable allenylindium species are stabilized and play an increased role in equilibrium (Scheme 13).

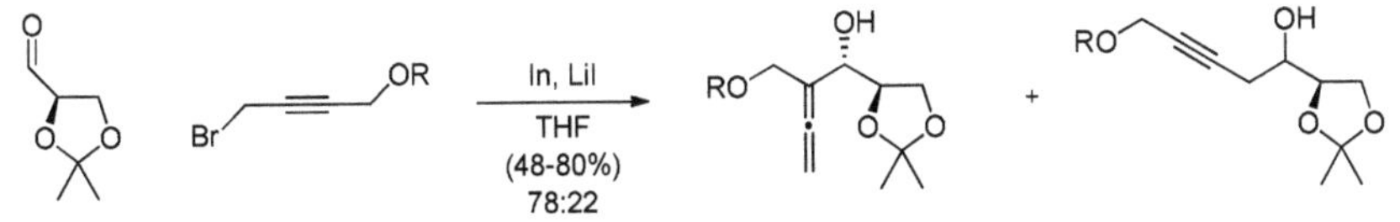

Scheme 13. Propargylation/allenylation of isopropylidene glyceraldehyde.

Subsequent ozonolysis and deprotection of allenyl alcohols furnished D-erythro-2-pentulose. The synthetic utility of the indium propargylation/allenylation was further demonstrated by subjecting the obtained substrates to various intramolecular gold catalyzed cyclizations[38] (Scheme 14).

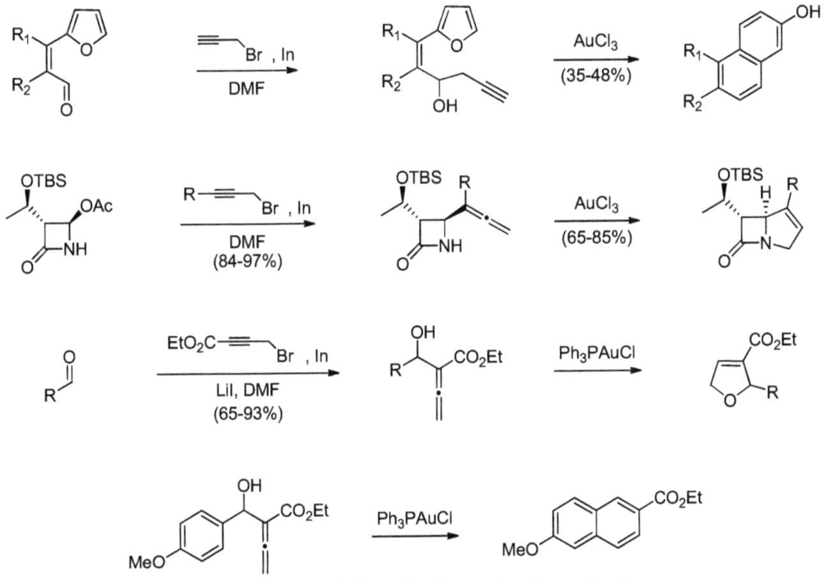

Scheme 14. Allenylation/propargylation followed by intramolecular cyclization.

Thus a rapid construction of highly functionalized frameworks was achieved, furnishing an array of heterocyclic compounds and naphthalene derivatives.

1.1.3 Indium mediated alkylation

Although the preparation of alkylindium reagents has been known for a long time[39] the use of these reagents in C-C bond formations has been scarce. The reason for this being on the one hand the very slow formation of the alkylindium species which requires the reactive Rieke indium3 (see section 1.1), on the other hand these reagents are incapable of reacting with carbonyl compounds due to their low reactivity. However, alkylindium reagents can be used in palladium catalyzed cross-couplings with aryl halides[40] (Scheme 15).

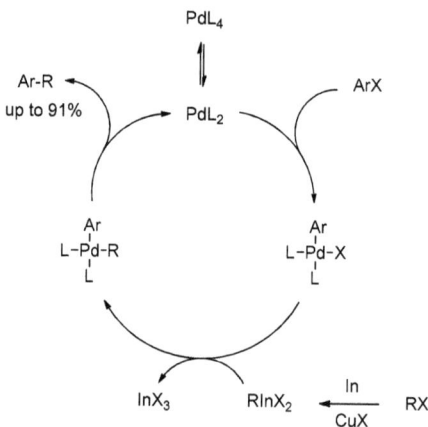

Scheme 15. Indium mediated cross-coupling of alkyl and aryl halides.

Owing to the mildness of alkylindium reagents, no competitive β-hydride eliminations or homocouplings are encountered and no protecting groups are required, which renders this reaction a complementary method to similar cross-couplings.

1.2 Titanium mediated aldol-type reactions in organic synthesis

Titanium reagents are used in a multitude of chemical transformations. In general, many of these reactions harness the strong Lewis acidity of titanium(IV) reagents, which promote C-C and C-O bond formations such as Mukaiyama and Evans aldol, Diels-Alder, Carbonyl-ene and the Sharpless epoxidation reaction among others. Furthermore (ansa-)titanocene complexes are used for enantioselective reduction processes, carbonyl olefinations and olefin polymerization reactions. The following sections will give an excerpt of the development of asymmetric aldol-type C-C bond formations throughout the past fifty years.

1.2.1 The Mukaiyama aldol reaction

Aldol additions use two carbonyl compounds to create β-hydroxy carbonyls, a general structural motif abundant in many natural products. Since there is a potential formation of multiple regio- and stereo isomers, it is of paramount importance to have mechanisms of control over both. The Mukaiyama aldol reaction (MAR) first provided a solution for these issues. Inspired by earlier work of Wittig[41] in 1963 on the enamide aldol reaction, in 1973 Mukaiyama and coworkers came up with the idea of using silyl enol ethers (SEE's) as stable metal enolates[42] for aldol additions (Scheme 16).

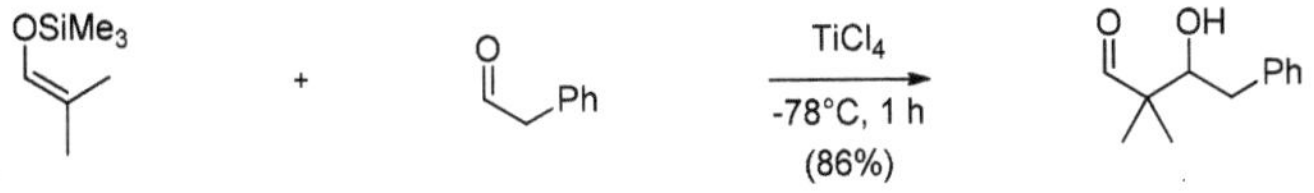

Scheme 16. MAR of benzyl and isopropyl aldehyde.

When activated with stoichiometric or catalytic amounts of $TiCl_4$, a strong metallic Lewis acid which was still easy to distill, these SEE's readily reacted with a range of different aldehydes and ketones. This not only solved the problem of self additions but also the regioselectivity could be controlled *via* the choice of base during the SEE preparation (Scheme 17). When a bulky base was used, it selectively deprotonated the less hindered side of a ketone. This observation was also confirmed by House *et al.*[43]

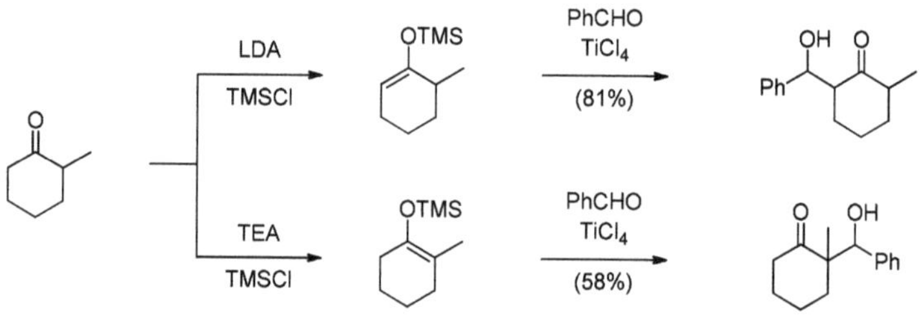

Scheme 17. Regio controll in the MAR.

Later on, also silyl ketene acetals[44] were used in the aldol addition followed by silyl dienol ethers[45] in the vinylogous MAR (Scheme 18), which is especially important in natural product synthesis[46].

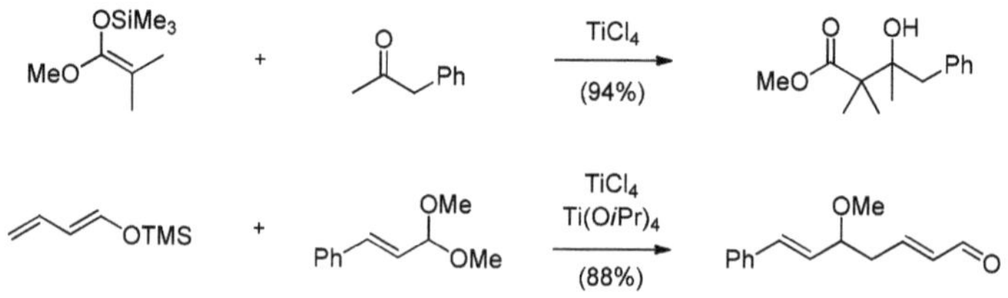

Scheme 18. Silyl ketene acetals and silyl dienol ethers in the MAR.

Unlike the respective lithium enolates, these compounds unfold their nucleophilicity in the γ-position, which can be rationalized in terms of larger orbital coefficients at this position. A stereochemical model for the MAR which assumes an open-chained transition-state (Scheme 19) and gained wide acceptance was first proposed by Noyori *et al*[47] in 1980.

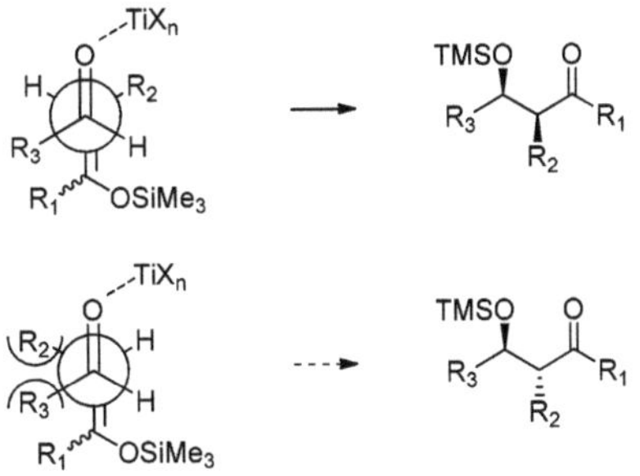

Scheme 19. Open-chained transition-state of the MAR.

The antiperiplanar alignment leading from both (*E*) and (*Z*) configured SEE's to *syn* products is assumed to be preferred as opposed to the unfavorable gauche conformation. Investigations concerning the control of stereochemistry were commenced in the 1980s. Evans *et al* developed an asymmetric boron-mediated aldol reaction by using chiral oxazolidinone auxiliaries (see section 1.2.2). Masamune, Paterson and Corey introduced chiral boron triflate Lewis acids[48] and Mukaiyama reported on the use of L-proline derived chelating diamine bases for enantioselective tin(II) mediated aldol additions[49] (Scheme 20).

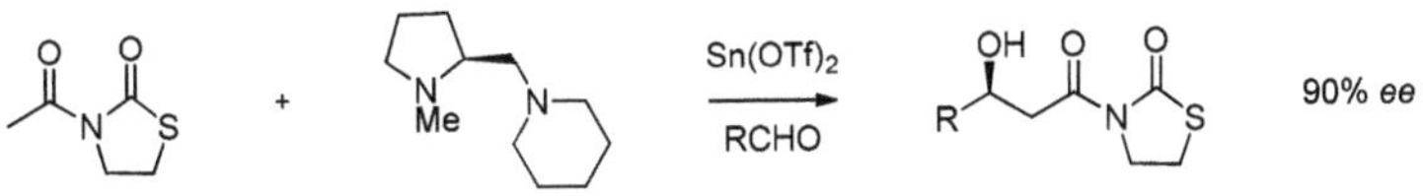

Scheme 20. Enantioselective aldol reaction by Mukaiyama.

The concept of chiral Lewis acids was further elaborated throughout the 1990s and reactions with a multitude of chiral catalysts with excellent enantioselectivities were introduced (Scheme 21).

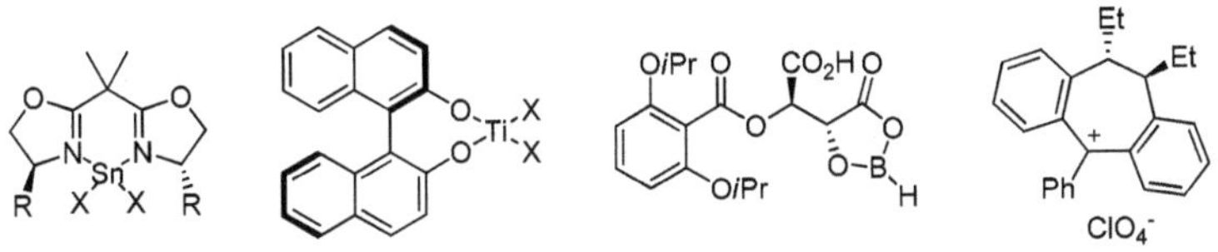

Scheme 21. Selected examples of chiral catalysts for enantioselective MAR.

Further important contributions in this field were made by Yamamoto and coworkers who developed a sequential MAR with 'supersilyl' enol ethers (Scheme 22)[50].

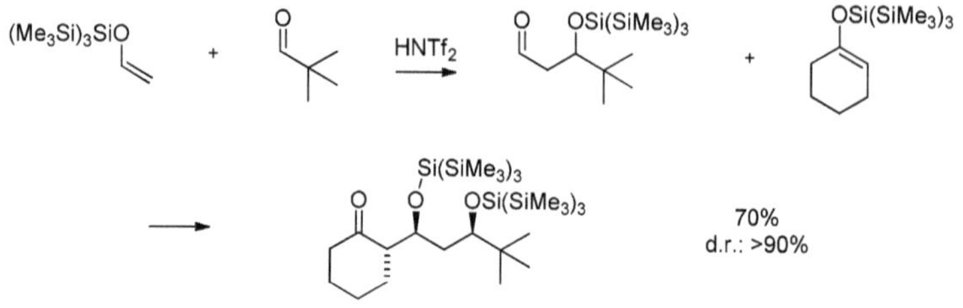

Scheme 22. Yamamoto's modification of the MAR.

This one-pot procedure uses only 0.05 mol% of $HNTf_2$ for activation and a very bulky silyl moiety, so that the intermediate aldehyde can be reacted with a further SEE, furnishing highly functionalized molecules in a single step. Concerning MAR's in natural product synthesis, the protocol of the Kobayashi group has to be mentioned[51]. This vinylogous MAR uses silyl ketene Evans auxiliaries which provided a hitherto unknown long range asymmetric induction with *anti* diastereoselectivity (Scheme 23). This procedure was used independently by Nicolaou and De Brabander for the total synthesis of palmerolide A[52].

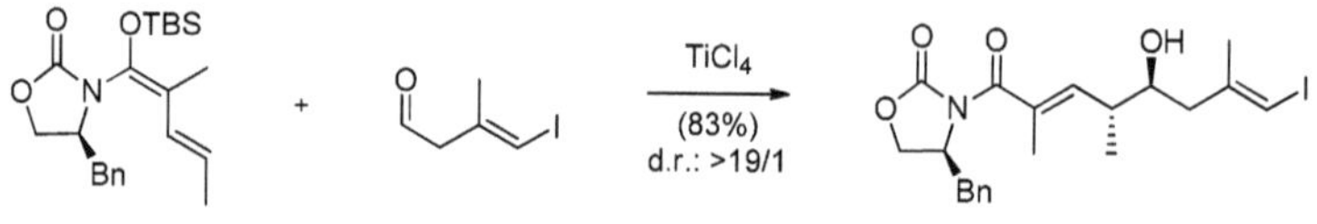

Scheme 23. Vinylogous MAR in the preparation of palmerolide A.

It was reasoned that the *anti* diastereoselectivity observed was due to unfavorable interactions within the *syn* transition-state, namely of the Lewis acid and the terminal methyl group, as well as the α-methyl group of the silyl ketene acetal and the aldehyde.

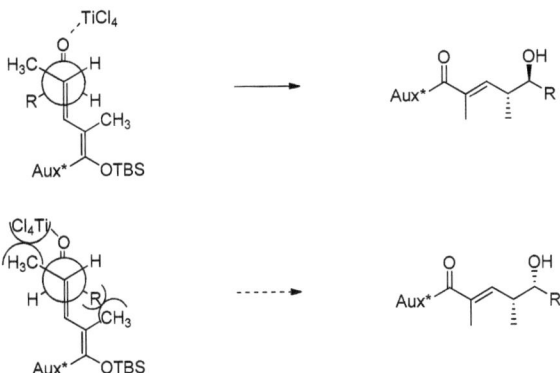

Scheme 24. Favored and disfavored transition-states in Kobayashi's vinylogous MAR.

Although Kobayashi's vinylogous MAR had shown to be a powerful tool for the construction of polyketides, one drawback remained the lack of a possibility to perform *syn* selective reactions. In 2009 Kobayashi[53] and Chen[54] reported on the use of chiral and achiral α- and β-heteroatom substituted aldehydes for the preparation of *syn* aldols (Scheme 25). The 1,3-dithianes first employed by Chen et al only displayed moderate selectivity but by continuing efforts it was found that *ortho* substituted benzaldehydes[55] also exhibited good *dr* values (Scheme 25).

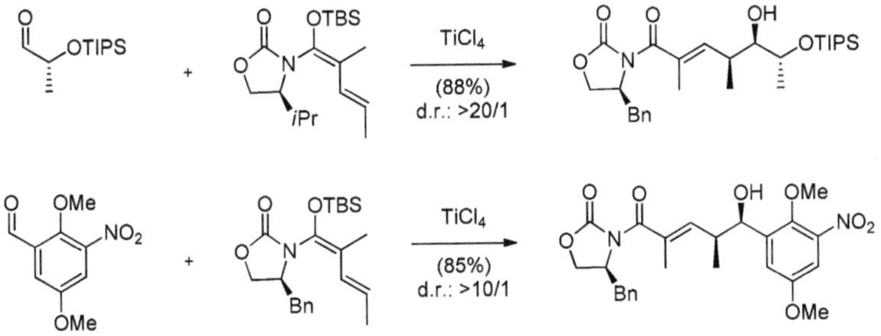

Scheme 25. α- and β-heteroatom substituted aldehydes in the *syn* selective vinylogous MAR.

It was reasoned that a transition-state must be involved which chelates the titanium Lewis acid between the heteroatom substituent and the carbonyl-O of the aldehyde and that the facial selectivity is inverted in order to avoid the unfavorable interaction of titanium and the terminal methyl group (Scheme 24). The facial selectivity of the reaction can also be changed by using excess $TiCl_4$[56]. Interestingly, under these conditions again *syn* products arise, even without chelating groups on the aldehyde. In 2012 Kalesse *et al* synthesized (Z) configured silyl ketene acetals which furnished *syn* aldols with non-chelating aldehydes and *anti* aldols with chelating ones[57] (Scheme 26).

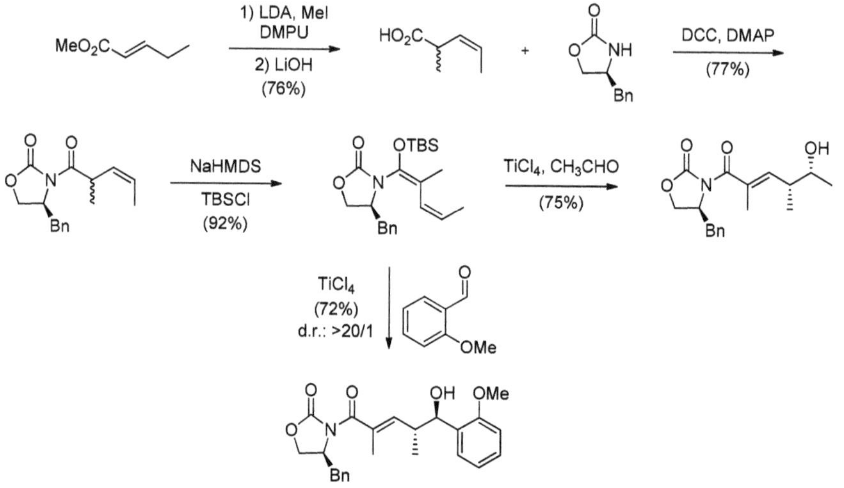

Scheme 26. Kalesse's (Z)-silyl ketene acetals in the preparation of *syn* non-chelate and *anti* chelate products.

This methodology completed the array of stereochemical control mechanisms in Kobayashi's vinylogous MAR and was further proof for the proposed transition-state. Dudley and coworkers used this protocol for a new formal total synthesis of palmerolide A[58] without the necessity to perform the stereochemical inversion applied by Nicolaou and De Brabander[52] (Figure 1).

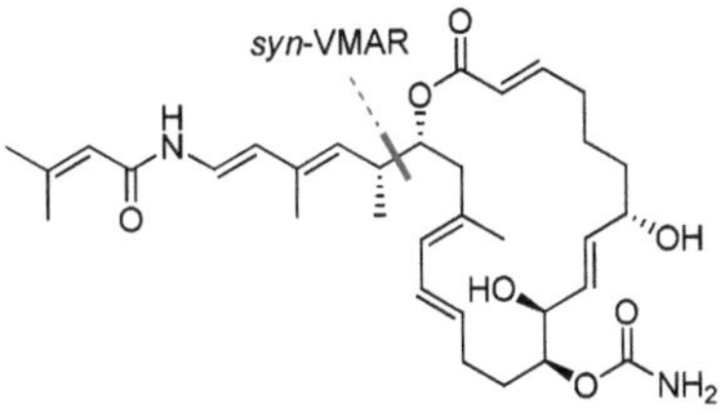

Figure 1. Palmerolide A.

The Mukaiyama aldol reaction has been investigated under constant improvement and expansion for more than 40 years now. It has become a pivotal transformation for the preparation of complex natural products such as polyketides and has taken its place among the fundamental tools of the synthetic organic chemist. Nevertheless it can be expected that further elaboration of this methodology will be performed in order to diversify the structural motifs available.

1.2.2 The Evans aldol reaction

A number of different mechanisms of control over regio- and stereoselectivity in aldol additions have been developed throughout the 20th century (see also section 1.2.1). In 1981 Evans *et al* reported on the stereochemical control of aldol additions by using boron Lewis acids[59]. At this time it was already known that distereoselection is partly defined by the geometry of the metal enolate[60] (Scheme 27).

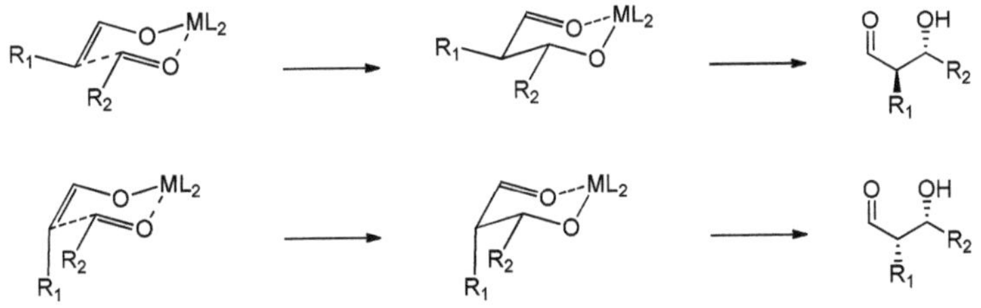

Scheme 27. Transition-states of (*E*) and (*Z*) metal enolates in aldol reactions.

It was assumed that *anti* aldol products were formed by pericyclic reactions of (*E*) configured enolates with carbonyls whereas (*Z*) enolates furnished the respective *syn* aldols, which is consistent with the Zimmerman-Traxler model[15]. With this preliminary knowledge, Evans and coworkers set out to investigate the stereoselective generation of boron enolates by modifying the metal ligands, the amine base and the solvent. Inspired by the work of Mukaiyama, dibutylboryl trifluoromethanesulfonate in combination with a tertiary amine base were used for the enolization of various ketones. The enolate species were subsequently trapped by double transmetalation with nBuLi followed by TMSCl for direct comparison with authentic samples. These studies gave further insight into the enolization mechanism and enolate equilibria. Under kinetic conditions, *anti* deprotonation with sterically hindered bases leading to (*Z*) enolates is preferred over *syn* deprotonation furnishing (*E*) enolates which is based on allylic strain arguments (Scheme 28).

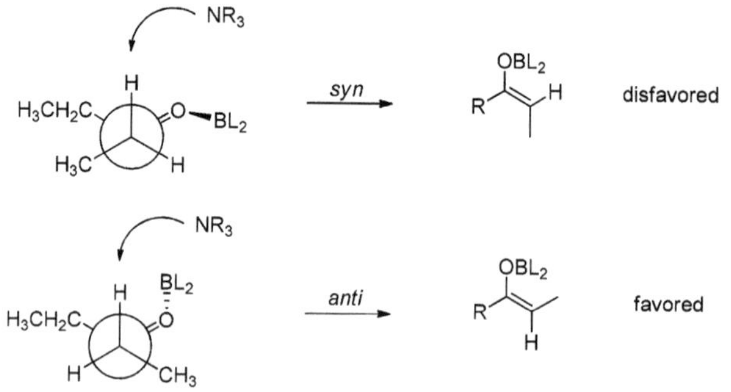

Scheme 28. Enolization of ketones, *syn* vs. *anti* deprotonation.

Thus, the use of Hünig's base (diisopropylethylamine) instead of 2,6-lutidine in the example above resulted in an increase in selectivity from 70:30 to >97:3. Additionally the exchange of n-butyl to cyclopentenyl ligands on boron could be used to fine-tune the selectivity and most importantly a consistent correlation between enolate geometry and aldol product stereochemistry was found. Furthermore the screening of different solvents showed that less polar ones such as dichloromethane in general enhanced the selectivity. This behavior was attributed to transition-state 'compression' which supposedly results in a stronger influence of steric parameters that dictate the diastereoselection. In some cases (*E*) enolates and their corresponding *anti* aldol products could selectively be prepared under thermodynamic conditions and/or the aid of bulky boron ligands and ketones. However no general procedure for an *anti* selective aldol addition using boron enolates could be realized (for a highly *anti* selective magnesium catalyzed reaction by Evans *et al* see Ref. 61). Apart from these detailed studies on the mechanism of enolization, the stereochemical induction of chiral enolates was investigated. N-tosylated proline derivatives in this respect showed good selectivity (*syn*/*anti* = 9/1, *syn*-a/*syn*-b > 97/3) and it was reasoned that two transition-states might be possible, one of them being strongly disfavored due to steric repulsion of the boron ligands and the large N-Ts moiety (Scheme 29). Another argumentation frequently given is that shielding of the *si*-face of the aldehyde by the auxiliary occurs which therefore directs the attack of the enolate from the *re*-face. Unfortunately, the chair-like Zimmerman-Traxler transition-state proved not to be a general concept in auxiliary-mediated aldol additions. In some cases boat-like or open-chained transition states[62] have been proposed in order to account for unexpected product distributions (*vide infra*) which suggests that a complete understanding of these processes is still missing.

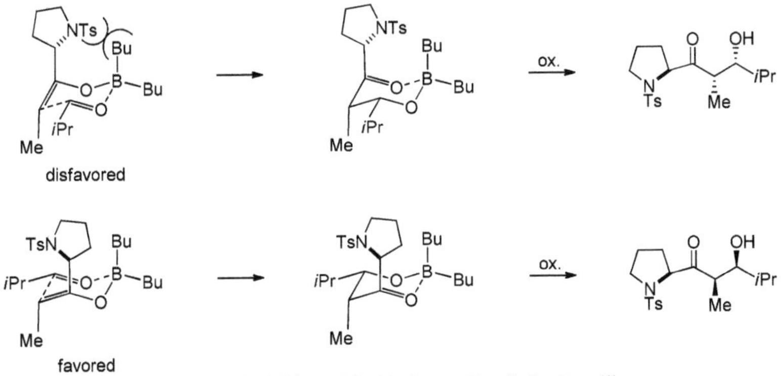

Scheme 29. Boron mediated aldol addition with chiral L-proline derived auxiliary.

In 1981, Evans and coworkers also reported on the use of L-valine and norephedrine derived oxazolidinone auxiliaries in their boron mediated aldol addition [63]. These very versatile compounds displayed even higher stereoselectivity (*syn/anti* = 250/1, *syn-a/syn-b* = 500/1) which in general produced less than 1% of combined unwanted diastereomers under the condition that R ≠ H (Figure 2).

Figure 2. Evans auxiliaries derived from L-valine and norephedrine.

However the acetate aldol reaction (R = H) could be achieved by using the respective –SMe substituted derivatives followed by reduction with Raney-Ni. Following this initial success Evans' 2-oxazolidinones shortly became one of the most popular auxiliaries for the asymmetric formation of C-C bonds[64] and the scope of this reaction could be demonstrated in a number of natural product syntheses[65]. The advantages of oxazolidinone auxiliaries can be summarized as follows. They are easily accessible starting from bulk chiral pool substances; they induce high levels of enantioselectivity, most of their aldol adducts are crystalline and they can be cleaved off and recycled easily. They are usually prepared from the parent amino alcohol and phosgene or diethyl carbonate followed by treatment with *n*-BuLi and acid chloride or anhydride or alternatively *via* a DCC amide coupling (Scheme 30). However various other approaches for the construction of the oxazolidinone framework have been devised[66], like palladium catalyzed carbonylations or carboxylations followed by internal Mitsunobu-type substitution (Scheme 30).

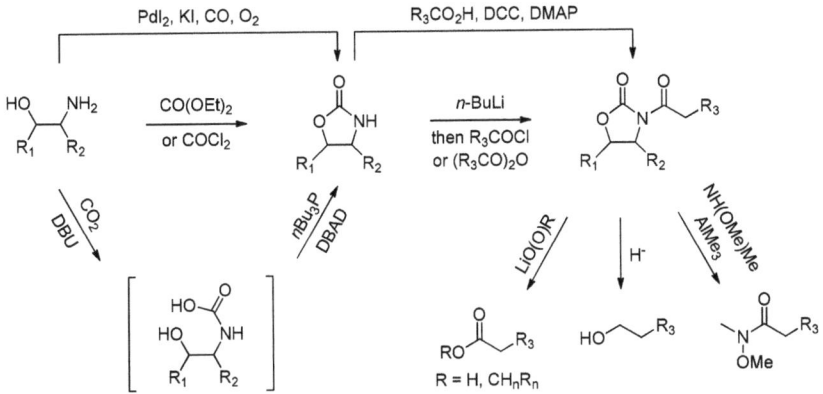

Scheme 30. Representative examples for the preparation of oxazolidinone auxiliaries.

The removal of the auxiliaries can be carried out either by simple saponification with hydro(pero)xide bases, trans-esterification with alcoholates[63], reduction with complex hydrides or by transamination with AlMe$_3$ and N,O-dimethylhydroxylamine (Scheme 30).[65a] Following the huge success of Evans' boron mediated asymmetric aldol reaction, the application of other metals such as tin[67] and titanium was investigated subsequently. Initially however titanium/DIPEA mediated reactions proved to be slightly less selective[68]. It was reasoned that titanium enolates lacking the stability and compactness of the respective boronates (B-O ~ 1.4 Å, Ti-O ~ 1.6 Å) participated in multiple reaction pathways[69] and therefore furnished lower overall selectivity. One possible competing pathway is illustrated in Scheme 31. After the loss of one chloride ion, titanium can coordinate to the carbonyl O of the auxiliary, thus changing the π-facial selectivity of the reaction. Crimmins and coworkers however were able to demonstrate that both 'Evans' and 'non-Evans' products were accessible by adjusting the reaction conditions[70]. This was achieved by using oxazolidinethione auxiliaries and by changing the stoichiometry of the Lewis acid and the nature of the amine base.

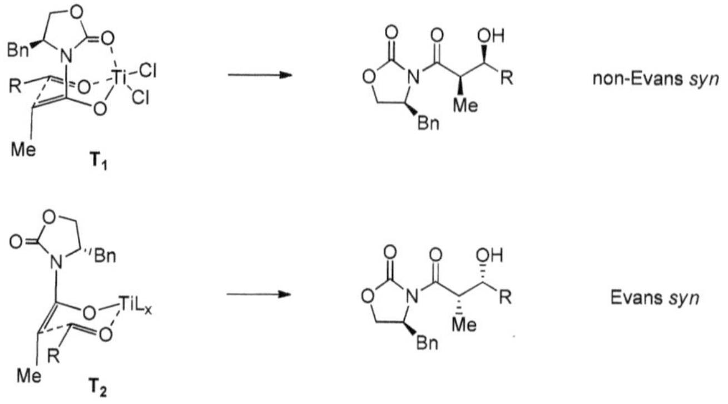

Scheme 31. Competing pathways in Ti mediated aldol additions.

It was found that oxazolidinethiones, beside their easier removal and recovery, also lead to enhanced selectivity owing to more stable and rigid titanium sulfur chelated enolates which furnished high diastereoselectivities of >98/2 in favor of the Evans *syn* product with 2.5 equiv of TMEDA as a base. DIPEA however gave inconsistent results. When (-)-spartein was used as a chiral base, no additional asymmetric induction but a reaction rate acceleration was observed improving isolated yields by more than 20%. Most interestingly, the diastereoselection could be inverted by changing the stoichiometry of the Lewis acid. When 2 equiv of TiCl$_4$ and 1 equiv of DIPEA were used, non-Evans *syn* products were obtained in an excess of >95/5. On the other hand Heathcock *et al* had reported on *anti* selective reactions with 2 equiv of (*n*Bu)$_2$BOTf and an acyclic transition state was proposed (Scheme 32)[71].

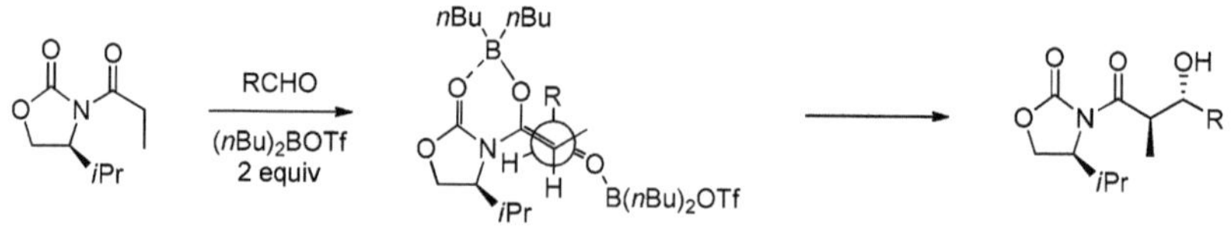

Scheme 32. Open chained transition state in the *anti* selective aldol addition by Heathcock *et al.*

Crimmins reasoned that chloride abstraction from the respective titanium enolate by the second equiv of TiCl$_4$ promotes the formation of non-Evans *syn* products *via* T_1 (Scheme 31). ^{1}H-NMR studies of the enolate species gave further support for this hypothesis. Two distinct enolates were identified, one of them being only found when excess $TiCl_4$ or an additional equiv of $Ag[SbF_6]$ was present to promote chloride abstraction. Thus, Crimmins *et al* were not only able to confirm the utility of titanium in asymmetric aldol reactions, but also a method for producing non-Evans *syn* aldols was developed which dispensed the need to prepare both enantiomers of the chiral auxiliary. Prior to these findings in 1993 Pridgen *et al* reported on a stereoselective Darzens reaction using oxazolidinone auxiliaries and different Lewis acids[72]. The Darzens reaction in principle represents an α-halogen variation of the aldol addition discussed in this section (Scheme 33).

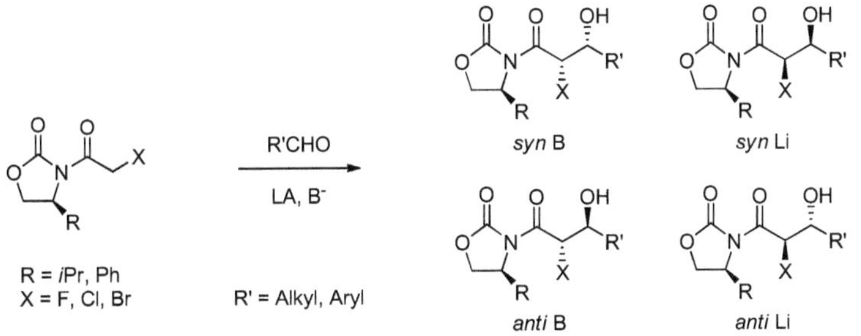

Scheme 33. Pridgen's α-halo variation of the Evans aldol addition.

It was shown that steric control could be exerted be the choice of the Lewis acid which subsequently allowed for the preparation of enantiomerically enriched α,β-epoxy esters. Although much more sophisticated methods for the asymmetric synthesis of epoxides have been developed since then (see for example section 2.3.2) this method still represents an interesting way for the introduction of fluorine. The obtained diastereomers were denoted by the metal that furnished them as the major product and fluorine, chlorine and bromine substituted auxiliaries were tested in the aldol addition with alkyl and aryl aldehydes. Concerning the nature of the metal two groups were defined: Non-chelating [B, Ti, Sn(II)] and chelating metals [Li, Zn, Sn(IV)], although it is evident that Ti can participate in both chelated and non-chelated transition-states (*vide supra*). In general, only boron, titanium and tin gave good diastereoselectivity, whereas Li and Zn gave moderate results (B-O ~ 1.4 Å, Ti-O ~ 1.6 Å vs Li/Zn-O ~ 2.0 Å). Also the role of the nature of the aldehyde was examined; aromatic aldehydes mainly furnished *anti* aldols whereas aliphatic ones[73] produced the usual *syn* products with 'chelating metals'. 'Non-chelating' metals furnished *syn* aldols, regardless of the aldehyde architecture. It was reasoned that the difference in energy between chair-like and boat-like transition states, which lead to *syn* and *anti* products respectively, is sufficiently small to be overcome by adjusting the reaction conditions. Thus, also exceptions from the 'rules' stated above were encountered. As a premise all reactions were assumed to proceed through (Z)-enolates and without epimerization at the α-carbon. In order to explain the formation of the *anti* products from aromatic aldehydes, a boat- or twist-boat-like TS was proposed (Scheme 34).

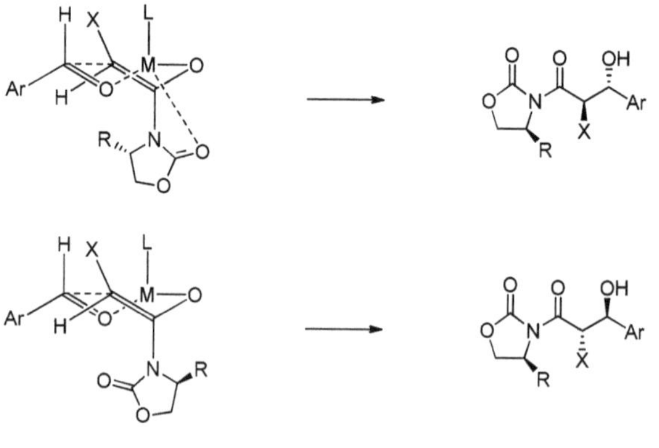

Scheme 34. Boat-like transition-states in the formation of *anti* aldols from (Z) enolates.

Concerning the influence of the nature of the halogen on the reaction, no definitive conclusions could be drawn, but it was found that fluorinated auxiliaries gave the best results in combination with TiCl$_4$ whereas (*n*Bu)$_2$BOTf proved to be ineffective. In conclusion, many approaches for the control of relative and absolute stereochemistry in asymmetric aldol reactions have been reported. Among these approaches, the use of chiral oxazolidinone auxiliaries, first exemplified by Evans *et al*, has enjoyed great popularity within the synthetic community. Although much effort has been devoted to establish a complete understanding of this reaction, a universal model might be elusive. However, it can be said that the reaction is governed by the following stereo-directing influences: The configuration of the chiral enolate (*E*) vs. (*Z*) and the reacting diastereo(enantio)topic faces of the enolate and the aldehyde, depending on their nature (chelating vs. non-chelating metals, aromatic vs. aliphatic aldehydes). The most commonly used Lewis acids for this type of reaction are (nBu)$_2$BOTf and TiCl$_4$ which both display remarkable selectivity. Concerning the cost and convenience of storage and handling, TiCl$_4$ is more advantageous compared to (nBu)$_2$BOTf which has to be kept under scrupulously anhydrous conditions or prepared freshly for best results.

1.2.3 Garner's aldehyde

Philip Garner first prepared his famous chiral building block from L-serine in 1984 (Scheme 35)[74]. It was found to be configurationally stable by Mosher ester analysis and subsequently used for the preparation of *threo*-β-hydroxy-L-glutamic acid.

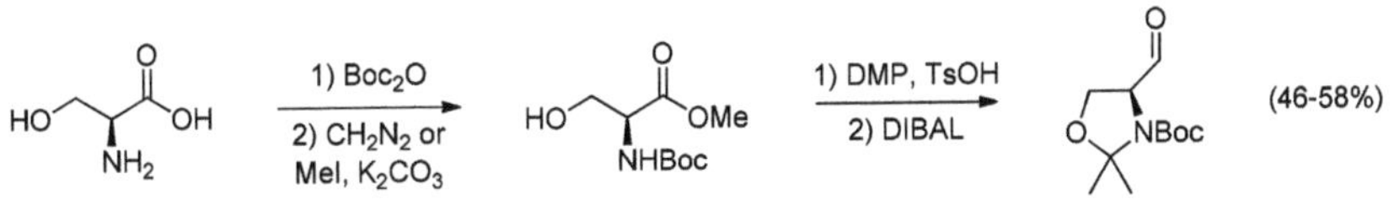

Scheme 35. First synthesis of 1,1-dimethylethyl 4-formyl-2,2-dimethyloxazolidine-3-carboxylate by Garner *et al*.

Three years later in 1987 Garner *et al* additionally reported on the preparation of the respective threonine derivatives of his aldehyde[75]. Since then, this original procedure has been subjected to many improvements, including the reversal of Boc protection and esterification[76], the use of BF$_3$.OEt$_2$ catalyst in the acetonide formation and replacement of the unreliable DIBAL reduction[77] with a LAH reduction, Swern oxidation sequence[78] (Scheme 36).

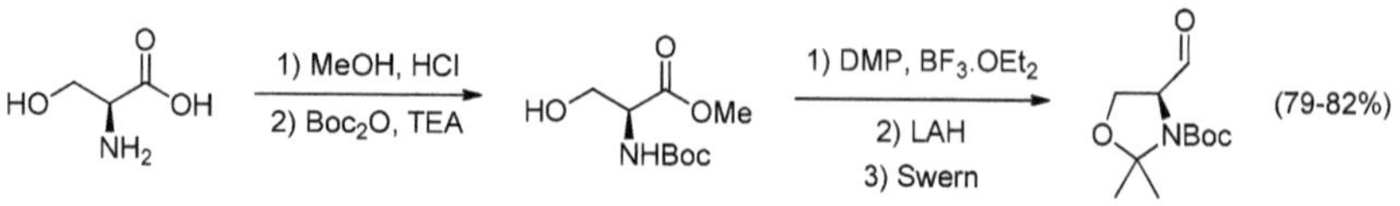

HO-CH(NH$_2$)-C(=O)-OH → 1) MeOH, HCl; 2) Boc$_2$O, TEA → HO-CH(NHBoc)-C(=O)-OMe → 1) DMP, BF$_3$·OEt$_2$; 2) LAH; 3) Swern → oxazolidine-NBoc-CHO (79-82%)

Scheme 36. Improved synthesis of Garner's aldehyde.

With these modified conditions, D- and L-Garner's aldehyde as well as the respective threonine derivatives could not only be prepared in very high yields, also the optical purity of the products was enhanced to >97% *ee* when DIPEA was used in the Swern oxidation (up from 93-95% *ee* reported by Garner). The addition of various nucleophiles to this aldehyde provides access to the 2-amino-1,3-dihydroxypropyl structural motif which is abundant in many natural products such as aminosugars[79], azasugars[80] and sphingosines[81]. The rigidity of the oxazolidine moiety prevents racemisation in the course of such reactions and methods for the selective preparation of both *threo* and *erythro* adducts using various organometallic reagents or aldol reactions have been developed[82]. However, the application of Evans' oxazolidinone auxiliaries in asymmetric aldol additions with Garner's aldehyde has been only scarcely investigated yet[83].

1.2.4 Aldol additions in carbohydrate synthesis

The most classic aldol addition in carbohydrate synthesis is the so called formose reaction. This oligomerisation of formaldehyde was first investigated by Loew and Fischer[84] who isolated fructose osazone from this complex mixture yielding reaction, which is considered to be the origin of RNA in prebiotic chemistry. Eschenmoser *et al* studied the reaction of glycol aldehyde phosphate with formaldehyde under 'primordial' conditions and observed the predominant formation of pentose diphosphates[85], which represented an important indicator for the significance of this reaction in the origin of ribonucleic acids. Further findings include the FeO(OH) catalyzed dimerisation of glyceraldehyde to form ketohexoses[86] and the propensity of such hydroxide minerals to absorb form-, glycol- and glyceraldehyde by reaction with immobilized sulfite which leads to a localized increase in concentration[87] of these RNA precursors.

An important methodology in carbohydrate synthesis is comprised by enzymatic aldol additions. Since enzymes can be considered perfect in terms of regio- and stereoselectivity, it is not surprising that the synthetic community increasingly promoted their application in recent years. In general, two types of aldolases are known. Type I is found in higher animals and plants and does not require cofactors, whereas type II is abundant in microorganisms and requires Zn^{2+} cofactor. X-ray structure analysis of RAMA (rabbit muscle aldolase) for example indicates that Lys-229 forms an enamine with dihydroxyacetone phosphate (DHAP) which can subsequently add to a natural or non-natural aldehyde substrate. (-)-1-deoxymannonojirimycin and (+)-1-deoxy-nojirimycin, two potent glycosidase inhibitors, are for example accessible in a three step synthesis with RAMA[88] (Scheme 37).

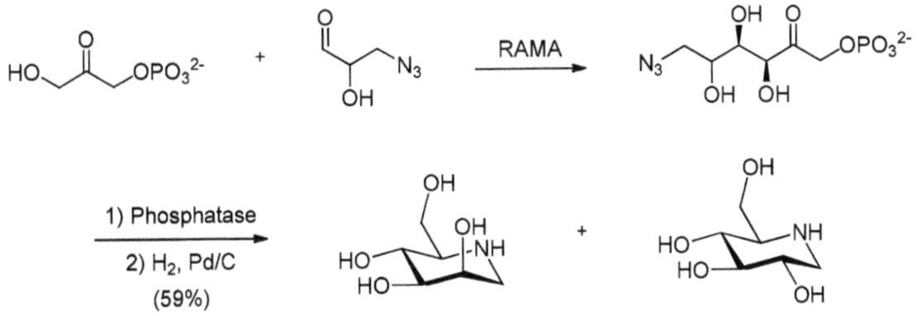

Scheme 37. RAMA catalyzed synthesis of imino sugars.

Racemic 3-azido-2-hydroxypropanal and DHAP under RAMA catalysis in this manner furnish azido-ketoses in a 4/1 ratio favoring the manno derivative. After phosphatase catalyzed removal of the phosphate group and reductive amination, the target imino sugars are generated. As expected, when enantiomerically pure 3-azido-2-hydroxypropanal is used, exclusive formation of the respective diastereomers is observed.

Concerning the chemical asymmetric synthesis of carbohydrates by means of aldol additions, three general approaches have to be considered. (1) Additions of enantiomercally pure aldehydes (glyceraldehyde, lactaldehyde) as a source of chiral information; (2) additions to chiral enolates (HYTRA, Evans' oxazolidinones) and (3) reactions using external chiral sources (chiral bases, lewis acids)[89]. A few representative examples for these approaches are given below. L-proline catalyzed reactions in this respect were originally investigated by Barbas and coworkers. This organocatalytic reaction was performed with phthalimido protected glycine aldehyde and isopropylidene dihydroxyacetone in order to prepare amino ketose derivatives, which were subsequently reduced to the corresponding alditols with L-Selectride (Scheme 38)[90].

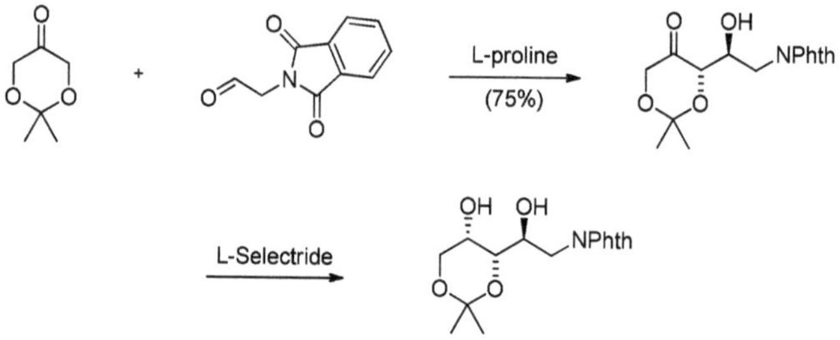

Scheme 38. Barbas' synthesis of amino alditols.

In a similar fashion, Enders et al applied Garner's aldehyde in the synthesis of 5-amino-L-psicose respectively –tagatose [91]. Additionally, asymmetric three component Mannich reactions with proline catalyst (Scheme 39) were investigated by List[92], Enders[93], Córdova[94] and Hayashi[95].

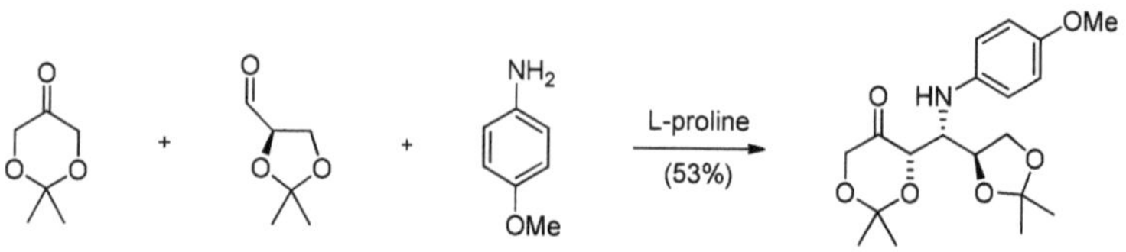

Scheme 39. Córdovas's asymmetric Mannich reaction; Synthesis of 4-amino-D-fructose.

This reaction furnished *syn*-products as opposed to the *anti*-products obtained in the corresponding aldol additions, which was explained by a steric repulsion between the aromatic ring and the proline catalyst. In general, both of these reactions are assumed to proceed *via* an enamine species, formed by the proline catalyst and the aldol acceptor (enolate equivalent), whereby the addition of the aldehyde occurs from the same face of the carboxylate moiety owing to hydrogen-bonding interactions (Figure 3).

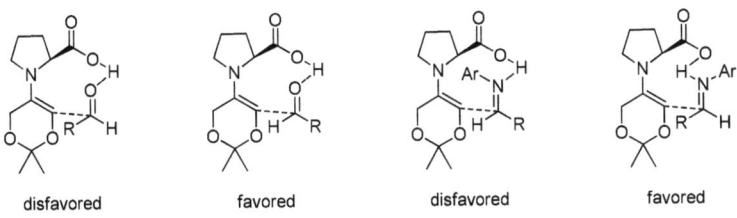

disfavored   favored   disfavored   favored

Figure 3. Possible transitions states in proline catalyzed aldol additions.

Furthermore, Mukaiyama and coworkers have demonstrated the utility of asymmetric aldol additions in carbohydrate synthesis by applying their chiral proline derived bases in the preparation of D-ribose[96] (Scheme 40).

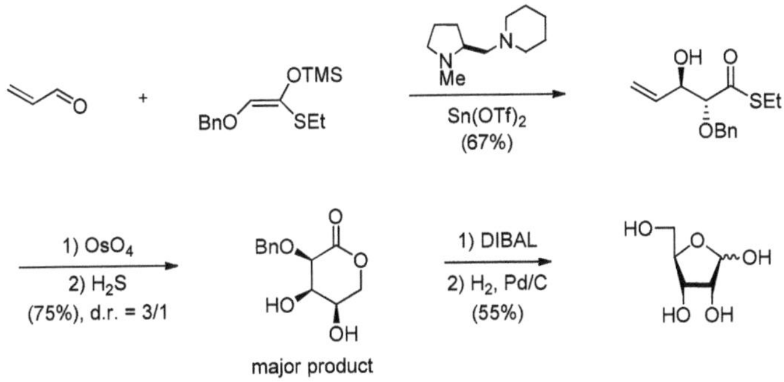

Scheme 40. Mukaiyama's synthesis of D-ribose with a chiral proline derived base.

A *de* and *ee* of more than 97% were achieved in this manner, although the osmium mediated dihydroxylation gave only moderate selectivity. In a similar fashion, L-fucose was later prepared by Kobayashi using crotonaldehyde as starting material[97].

An interesting approach for the synthesis of D-digitoxose from lactaldehyde was published by Braun et al in 1991[98]. This carbohydrate is a constituent of the digitalis glycosides found for example in foxglove (*digitalis purpurea*). For the preparation of this 2,6-dideoxy-D-allose derivative Braun and coworkers applied (*R*)-(+)-2-hydroxy-1,2,2-triphenylethyl acetate (HYTRA) as a chiral enolate in an aldol addition with protected 4-deoxy-D-erythrose (Scheme 41).

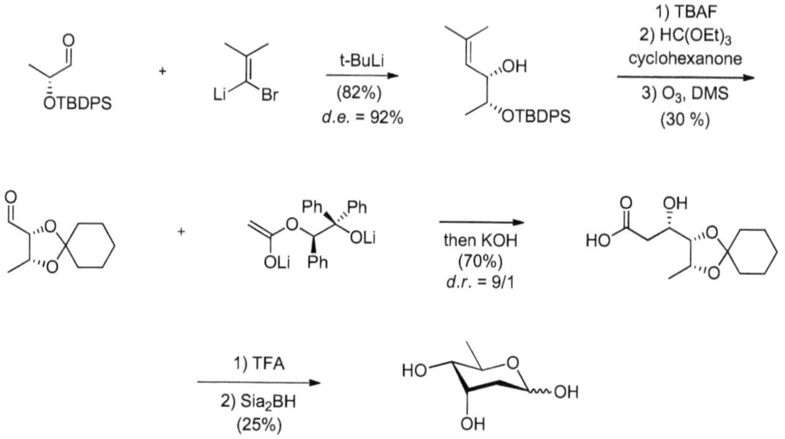

Scheme 41. Braun's synthesis of D-digitoxose.

The application of **HYTRA** represents an efficient way of performing an asymmetric 'acetate-aldol' reaction, which is not easily achieved with the corresponding oxazolidinone auxiliaries, developed by Evans et al (see section 1.2.2). An impressive example concerning the application of Evans' auxiliaries in natural product synthesis was published in 1992. The first total synthesis of calyculin A[99], a toxin found in marine sponges (*Disodermia calyx*), was performed by mainly enolate-based bond formation, thus constructing 10 of the 15 stereocenters. The C_{33}-C_{37} fragment of calyculin A contains a 2-amino-1,3-dihydroxypropyl structural motif which was constructed by two consecutive *syn* respectively *anti* selective aldol additions (Scheme 42).

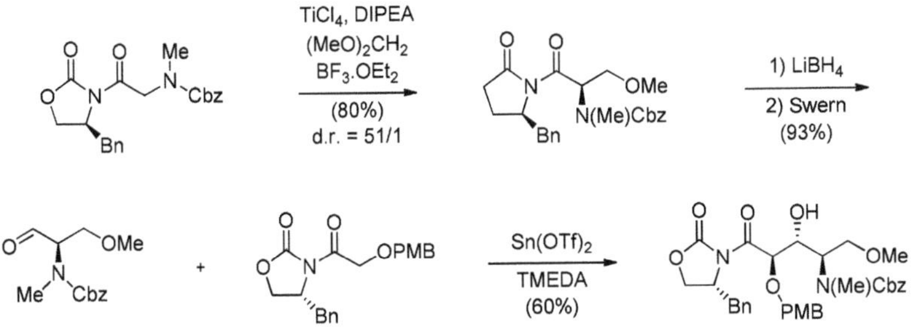

Scheme 42. Synthesis of the C_{33}-C_{37} fragment of (+)-calyculin A by Evans *et al.*

Although a total amount of 24% of other diastereomers was obtained after the tin(II) mediated aldolization, the 4-amino-D-ribonic acid fragment of calyculin A (Figure 4) was successfully obtained in this manner. This natural product was found to be a highly potent serine/threonine protein phosphatase inhibitor (IC_{50} for PP1 = 2 nM) which induces contraction of smooth muscle fibres and promotes tumor growth.

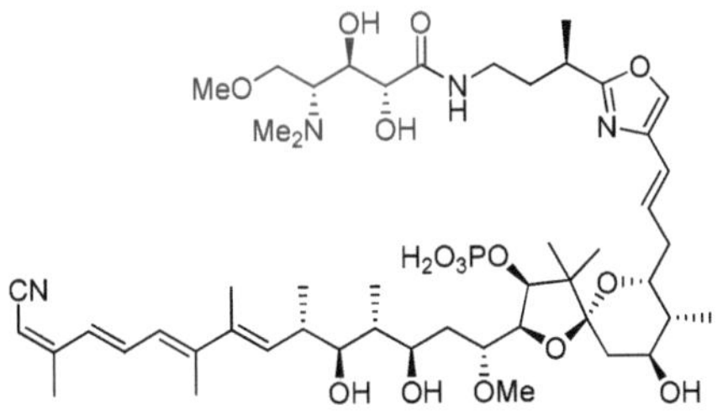

Figure 4. (+)-Calyculin A.

In summary, asymmetric aldol additions comprise highly useful tools in carbohydrate synthesis owing to the simultaneous and selective construction of vicinal hetero-substituted stereocenters. A large number of approaches have been devised in order to meet the requirements of the individual substrates. Thus, rare and unnatural compounds are accessible in short, straight forward synthetic sequences. Although the application of chiral 'additives' in this manner might seem atom inefficient, most of these compounds are easily prepared from bulk chemicals and can be recycled in many cases.

2 Results and discussion

2.1 Aim and background of the projects

This thesis comprises two projects; on the one hand, a new approach for the synthesis of 2-acetamido-heptoses and octoses has been devised and on the other hand, a versatile methodology for the preparation of amino-fluoro functionalized pentoses and hexoses was established. The idea for the first project was to extend the known indium mediated allylation of carbohydrates[100] by additionally applying an epoxidation, azide opening sequence for the introduction of nitrogen (Scheme 43).

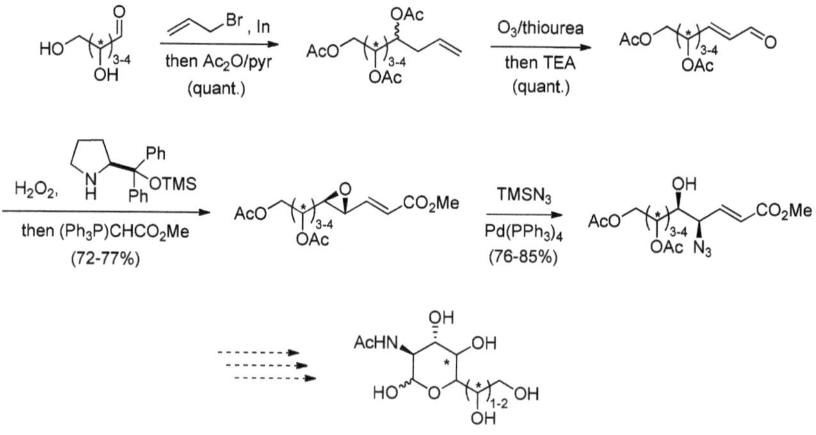

Scheme 43. Synthesis of 2-acetamido-heptoses and octoses.

Concerning the functionalisation of the carbon chain, the optimization and detailed discussion of the reactions has been the scope of previous theses.[101] The scope of this thesis was to develop a suitable deprotection protocol for the allylic azides obtained, which hitherto had not been achieved.

The second project, the preparation of amino-fluoro functionalized carbohydrates, was initiated based on a paper about a one-pot epoxidation/fluoride opening protocol for allylic amines published in 2012 by

Davies et al[102]. To demonstrate the synthetic utility of this reaction, Garner's aldehyde was transformed to the corresponding fluorinated phytosphingosine (Scheme 44).

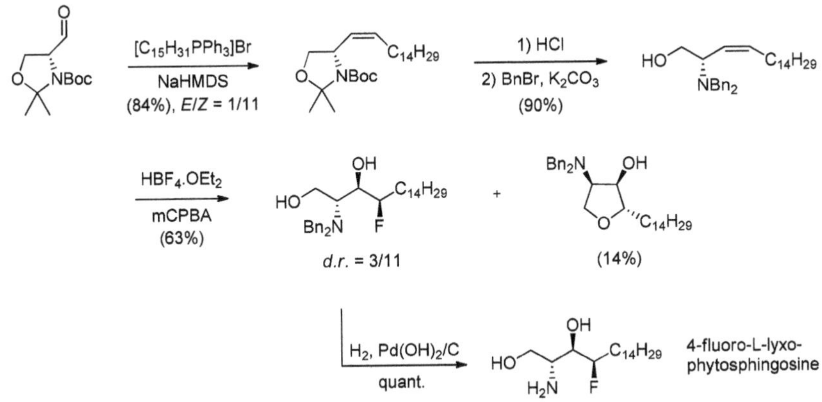

Scheme 44. One-pot epoxidation/fluoride opening protocol in the preparation of fluorinated phytosphingosines.

Additionally, the diastereoselection could be inverted by reducing the amount of HBF$_4$.OEt$_2$ and mCPBA used, which induced the predominant formation of the corresponding all-*syn* product. It was also found that tetrahydrofurane side products were formed in the course of the reaction owing to intramolecular epoxide opening by the free hydroxyl moiety. Nevertheless, we were interested whether it would be possible to perform the same reaction sequence using a stabilized Wittig ylide in the olefination step in order to prepare 4-amino-2-fluoro pentoses. Additionally, two alternative approaches were envisioned. A Barbier-type allylation of Garner's aldehyde followed by ozonolysis and electrophilic α-fluorination and an aldol-type addition of fluoroacetyl oxazolidinone chiral auxiliaries (Scheme 45).

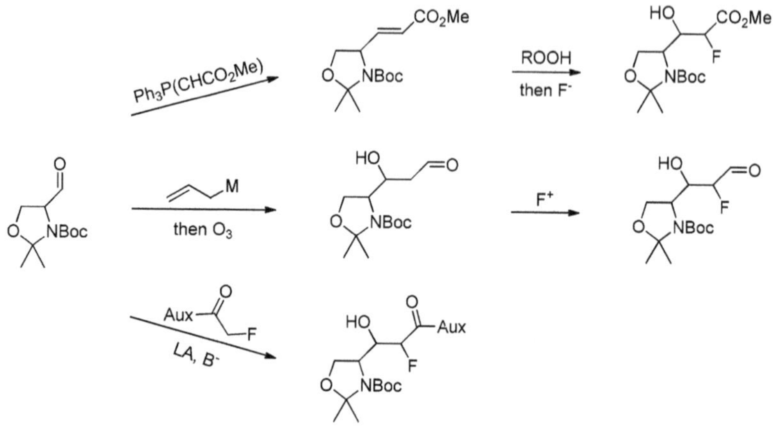

Scheme 45. Outline for the incorporation of fluorine.

The second approach was based on publications by Jørgenson [103] and MacMillan [104] et al about the stereoselective α-fluorination of aldehydes by combining iminium catalysis with an F⁺ source such as N-fluorobenzenesulfonimide (NFSI) (see section 2.4.2). The deoxy-aldehyde substrates for this reaction in turn should be easily accessible by metal mediated allylation, applying literature procedures [105]. The third approach mentioned above should harness the stereoselective Darzens reaction by Pridgen et al[72], thus forming both stereocenters in one step. All of the proposed approaches are straight forward and should provide an easy access to the class of 4-amino-2,4-dideoxy-2-fluoro-pentoses and additionally, 4-amino-2,4,6-trideoxy-2-fluoro-hexoses by applying the corresponding threonine derived aldehydes.

2.2 Motivation for the projects

2.2.1 Motivation for the synthesis of higher amino sugars

Higher analogues of amino sugars represent an interesting, biologically active class of substances which is not very well investigated yet. Although not many naturally abundant derivatives are known, two interesting examples include the aminoglycoside antibiotics apramycin[106] and destomycin[107] (Figure 5) which are both used in veterinary medicine.

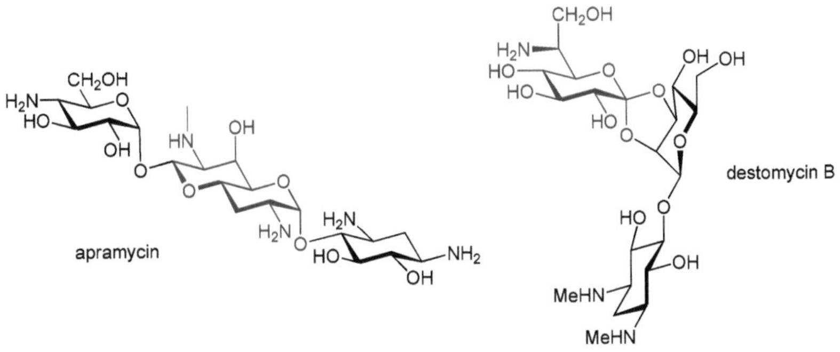

Figure 5. Amino-heptose and –octose containing antibiotics apramycin and destomycin B.

Even though these compounds are industrially produced by fermentation, one has to keep in mind that biotechnological production methods are not flawless. Concerning the purity of products, chemical synthesis is still superior and the feasibility of preparing comprehensive substance libraries is especially important since there is an ever increasing demand for new antibiotics owing to the evolution of resistant bacterial strains[108]. Therefore it may be desirable in the future to combine biotechnology and organic synthesis by means of introducing synthetic amino-sugars into cell-free or whole cell systems in order to produce structurally diverse biologically active aminoglycosides. The classic approach for the preparation of higher amino-sugars is the amino-nitrile synthesis. Originally applied by Kuhn and Kirschenlohr[109] in the synthesis of glucos- and galactosamine, Perez et al[110] used the amino-nitrile protocol for the preparation of amino-heptoses (Scheme 46).

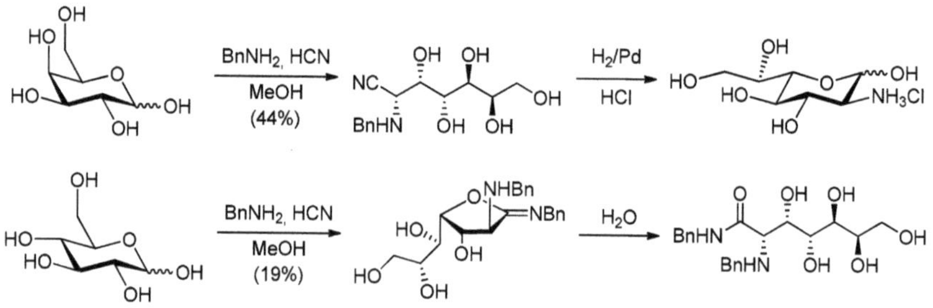

Scheme 46. Scope and limitations of the amino-nitrile synthesis.

Unfortunately, this reaction suffers from general low reproducibility and yield owing to side reactions, one of which has been identified to furnish amide products with a second equivalent of amine (Scheme 46). In the case of galactose, the aldononitrile product crystallizes immediately from the reaction mixture; therefore D-glycero-L-gluco(manno?)-heptosamine can be prepared in this manner. A more modern and reliable strategy for the preparation of (amino)-heptoses and octoses found frequently in the literature consists of an oxidation/Wittig-type chain elongation/dihydroxylation sequence (Scheme 47).

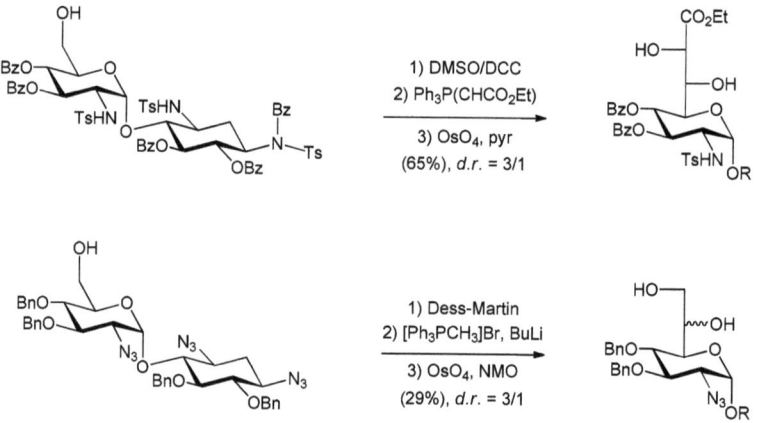

Scheme 47. Synthesis of apramycin[111] and paromamine derivatives[112] by an oxidation/Wittig-chain elongation/ dihydroxylation sequence.

Our methodology in comparison has some major advantages: (1) Control of the stereochemistry by application of a chiral epoxidation catalyst and therefore the feasibility of preparing all different isomers of 2- or 8- and 7-amino-heptoses and octoses, (2) shortening of synthetic sequences by reducing the amount of protecting group chemistry and (3) high overall yields due to early functionalization of the precursors. In summary, our approach should provide a stereoselective access to the substance class of higher amino sugars by using modern organometallic chemistry and organocatalysis.

2.2.2 Motivation for the synthesis of fluorinated amino sugars

Various rare 2- and/or 4-amino-6-deoxy functionalized hexoses and pentoses are located in the lipopolysaccharides (LPS's) of the outer cell wall of Gram-negative bacteria[113] (Figure 6).

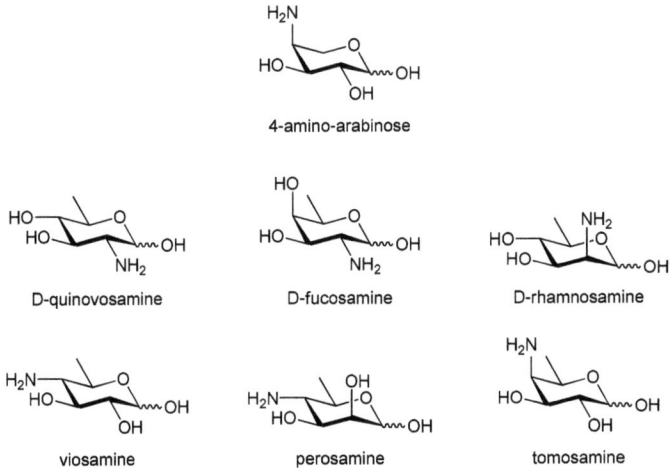

Figure 6. Selection of 4-amino-pentoses and 2- and 4-amino-6-deoxy-hexoses found in bacteria; compounds with stereo-descriptors are found in both D- and L-form.

These compounds are involved in immune response, pathogenicity, and adaptation mechanisms such as antibiotics resistance[114]. The O-antigen of the bacterium *vibrio cholerae* for example is composed of perosamine repeating units (Figure 7).

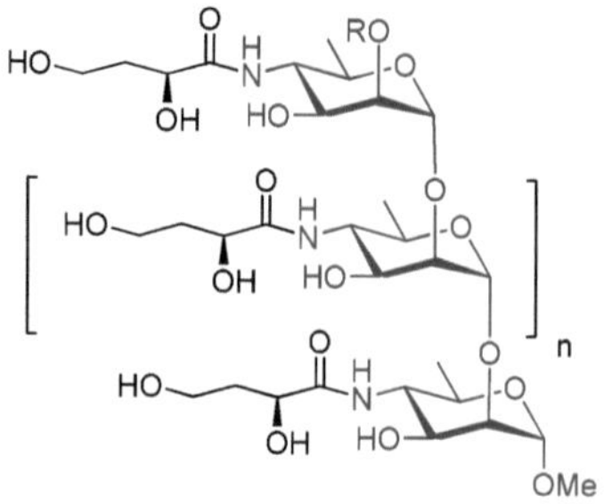

R = H, Inaba serotype
R = Me, Ogawa serotype

Figure 7. O-antigens in common serotypes of *vibrio cholerae* O:1.

The preparation of fluorinated analogues of these compounds is interesting within two aspects: (1) the elucidation of enzyme mechanisms and binding aspects of antigen-antibody adducts[115] and (2) the preparation of anti-microbial agents[116] and vaccines for the treatment of infections and cancer. Owing to the unique properties of the C-F bond[117], enhanced electrostatic interactions and different enzymatic reaction pathways may arise and lead to enzyme inhibition. Additionally, fluorine represents a bioisostere of hydroxyl moieties but cannot act as a hydrogen donor, which allows the localization of critical hydrogen bonding interactions with biomolecules[115]. Therefore, the decrease or increase of affinity constants towards fluorinated analogues maps the electronic environment of binding interfaces. Additionally, the 100% abundant ^{19}F nucleus allows for the application of NMR based investigations[118]. It has a sensitivity comparable to the proton, a large chemical shift range (-270 to +150 ppm for organic compounds) and large heteronuclear coupling constants ($^3J_{H,F}$ ~ 10-30 Hz, $^1J_{C,F}$ ~ 160-180 Hz for aliphatic compounds). For the preparation of fluorinated carbohydrates two popular strategies can be found in the literature. (1) The nucleophilic substitution of OH groups with sulfur-flouride reagents such as DAST[118] (Scheme 48) and (2) the electrophilic fluorination of glycals with F⁺ sources such as Selectfluor[119] (Scheme 49).

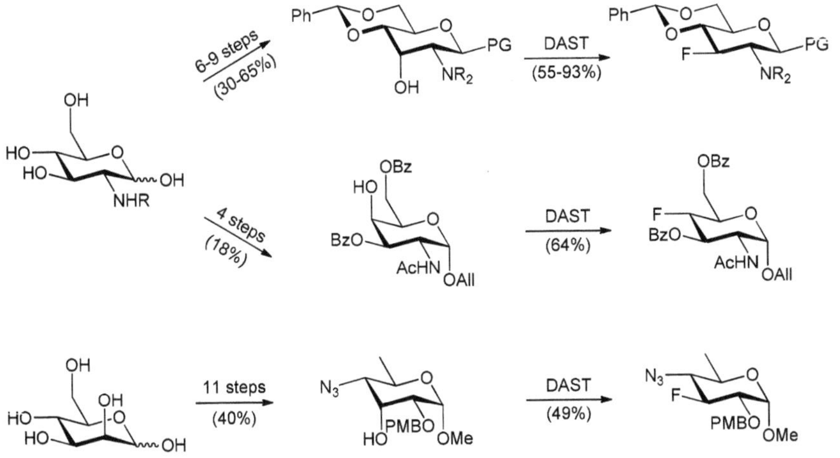

Scheme 48. Nucleophilic fluorination of protected amino sugars with DAST.

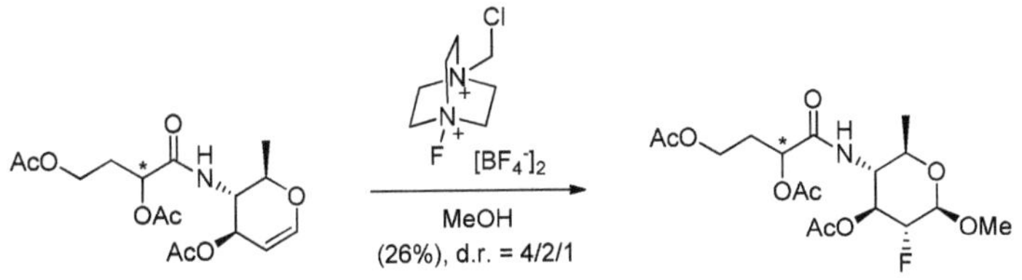

Scheme 49. Electrophilic fluorination of glycals for the preparation of cholera antigen derivatives.

All these approaches require extensive protecting group manipulations, which leads to long linear synthetic routes and low overall yields, especially if the amino- and 6-deoxy functionalities have to be introduced additionally. Another issue of the DAST reagent, despite its tendency to decompose explosively at higher temperatures, is the moderate yield sometimes obtained, which is especially problematic since it is mostly applied at a very late stage of the synthesis. Also the fluorination of glycals can be difficult as seen in the example above, where only low diastereoselectivity and yield were achieved. With this in mind, we set out to develop a synthetic route which should be much shorter and easier (Scheme 45) and which should provide a general access to the highly interesting class of 4-amino-2,4,6-trideoxy-2-fluoro-hexoses.

2.3 Synthesis of higher amino sugars

2.3.1 Indium mediated allylation of unprotected carbohydrates

We commenced our reaction sequence by using D-arabinose **1a**, D-galactose **1b** and D-glucose **1c** as starting materials in the indium mediated allylation which was performed under ultrasonication at 20-55 °C over 3-7 h either in EtOH/H$_2$O = 4/1 in the case of **1a** or EtOH/HCl 0.1 M = 4/1 in the case of **1b** and **1c** (Scheme 50). The stoichiometry applied was **1a-c**/indium metal (powdered)/allyl bromide = 2/4/7. Obtained yields were essentially quantitative, except for gluco-derivative **2c**, which was isolated in 70% yield after separation of non-allylated, peracetylated glucose.

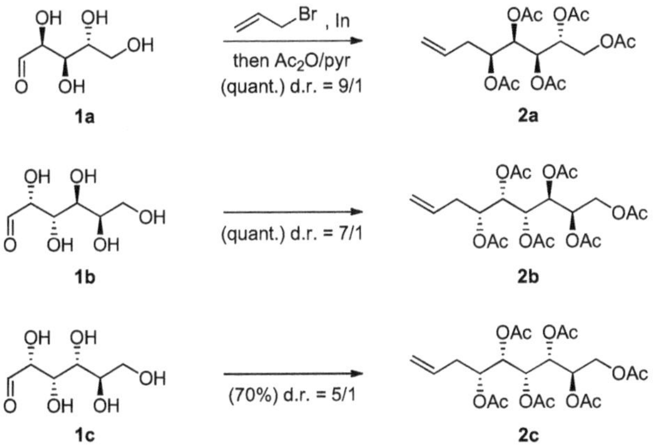

Scheme 50. Indium mediated allylation of **1a-c**.

After exhaustive acetylation of the diastereomeric mixtures obtained, ozonolysis was performed, followed by elimination with TEA, furnishing unsaturated aldehydes **3a-c** quantitatively (Scheme 51).

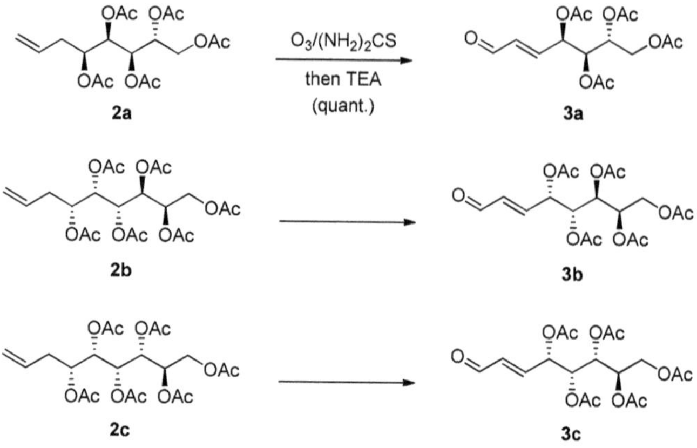

Scheme 51. Ozonolysis and elimination of compounds **2a-c**.

For convenience, thiourea was used for quenching of the ozonolysis, since it could be removed afterwards by simple filtration. The reaction mixtures were then directly treated with TEA for 30-50 min to furnish essentially pure compounds **3a-c** without any traces of (Z) isomers.

2.3.2 Epoxidation of unsaturated aldehydes

For the stereoselective epoxidation step an L-proline derived amine catalyst was used and the aldehyde moiety was subsequently masked *via* a Wittig olefination (Scheme 52), since the direct functionalisation of the corresponding α,β-epoxy-aldehydes was unsuccessful[101a].

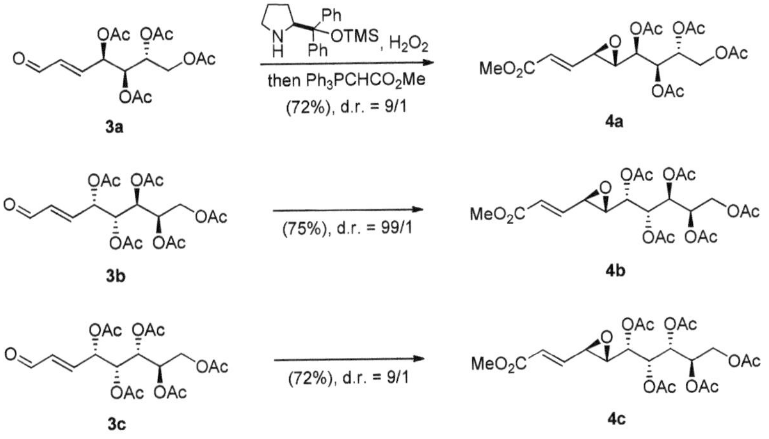

Scheme 52. Stereoselective epoxidation of compounds **3a-c**.

The organocatalytic epoxidation reaction, originally published by Jørgenson *et al*[120] in 2005, was performed in DCM with an aqueous hydrogen peroxide solution (50%) at -20 °C for 16 h and a catalyst loading of 15 mol%. The crude α,β-epoxy-aldehydes obtained were then directly treated with a stabilized Wittig ylide to furnish compounds **4a-c**. The catalytic cycle proceeds through a *trans*-iminium ion, formed by the amine catalyst and aldehyde, which is subsequently attacked by hydrogen peroxide in a conjugate fashion (Scheme 53). The bulky diphenyl(trimethylsilyloxy)methyl moiety in this case shields the *re* face of the substrate, therefore directing the nucleophilic attack from the *si* face.

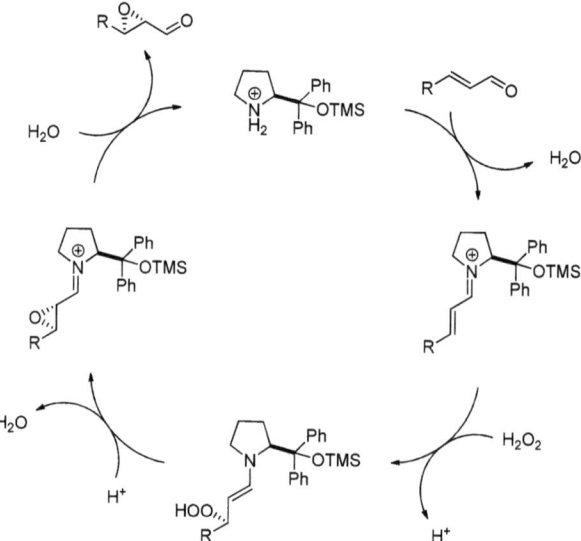

Scheme 53. Catalytic cycle of Jørgenson's epoxidation of α,β-unsaturated aldehydes.

After closing of the oxirane ring, the catalyst is hydrolyzed off, releasing the epoxide-products. Various azide sources and Lewis acids were screened in order to achieve the opening of these labile epoxides but unfortunately all attempts led to their decomposition. However, α,β-unsaturated-γ,δ-epoxy-esters **4a-c** could be opened cleanly by applying palladium chemistry.

2.3.3 Nucleophilic azide opening of epoxides

A palladium catalyzed, Tsuji-Trost type epoxide opening[121] of compounds **4a-c** was subsequently performed (Scheme 54), furnishing compounds **5a-c** which already feature the fully functionalized carbon skeletons of the desired 2-acetamido-heptoses and octoses.

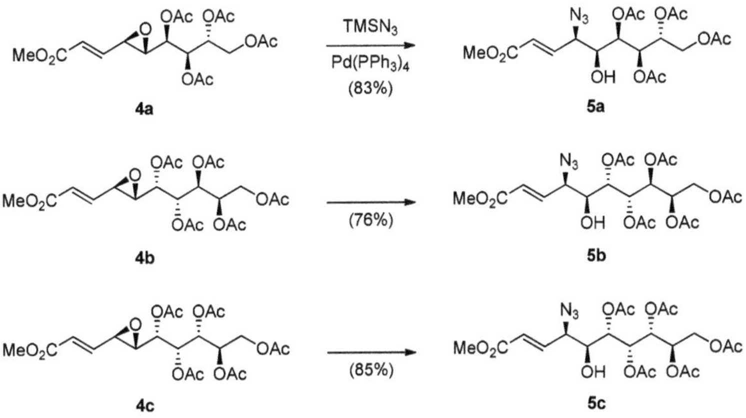

Scheme 54. Palladium catalyzed epoxide opening of compounds **4a-c**.

The azide opening of compounds **4a-c** was performed in THF for 1 h at room temperature under scrupulously inert conditions with 10 mol% of palladium catalyst. For best results, the use of fresh reagents of best quality and the successive addition of TMSN$_3$ and Pd(PPh$_3$)$_4$ to the substrate in solution proved to be crucial. The reaction occurs under net retention of configuration, which can be rationalized from the proposed catalytic cycle[121] (Scheme 55).

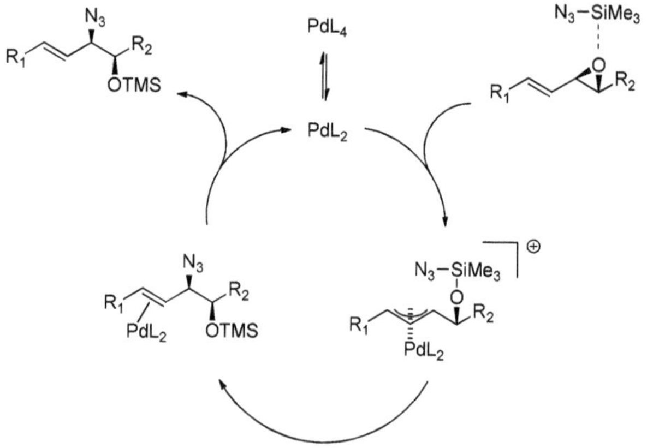

Scheme 55. Azide opening of allylic epoxides; catalytic cycle.

In the first step, the 18 valence electron palladium complex dissociates two of its PPh$_3$ ligands, forming the corresponding reactive 14 ve complex. The Tsuji-Trost reaction[122] then requires an allylic system with a leaving group, in this case an epoxide, so that in the next step oxidative addition of palladium under inversion of configuration and formation of a π-allyl complex occurs. This reaction is facilitated by the silicon, which receives the negative charge of the oxygen leaving group and bridges it with the azide nucleophile. This behavior also explains the regiospecificity of the reaction. In fact no S$_N$2' products are encountered, which would necessitate a 7-membered transition state as opposed to a 5-membered one for the corresponding S$_N$2 product. In the last step, the azide nucleophile attacks at the allylic position under inversion of configuration and palladium is reductively eliminated furnishing *syn* azido alcohols after work-up. In our hands, one problem remained to be the up-scaling of this reaction. When performed on a scale above 0.3 mmol, the yields usually dropped by about 20%, which forced us to perform the reaction in multiple small batches. Additionally, mixtures of acetate migration products were obtained in the case of compound **5a**, possibly owing to the *syn* relationship between C-5-OH and C-6-OAc (Figure 8).

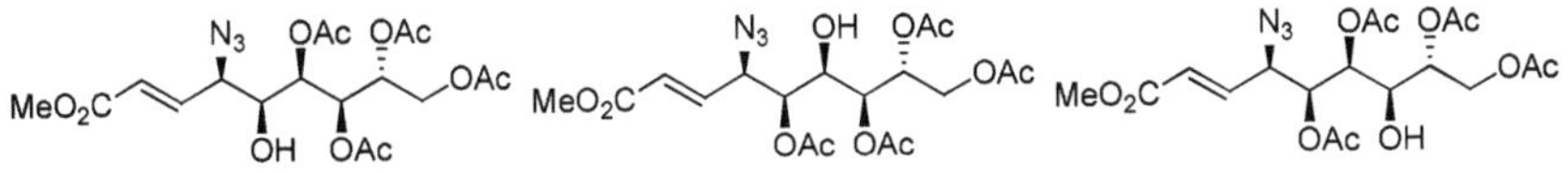

Figure 8. Products arising from acetate migration in compound **5a**.

2.3.4 Deprotection protocol

The main part of this project concerning the thesis at hand was to develop a short and easy deprotection sequence for compounds **4a-c**. It was already known, that these compounds were labile under basic conditions (prone to eliminations) and that acidic conditions promoted intramolecular Michael additions, forming tetrahydrofurane derivatives of the C-glycoside type (Scheme 56).

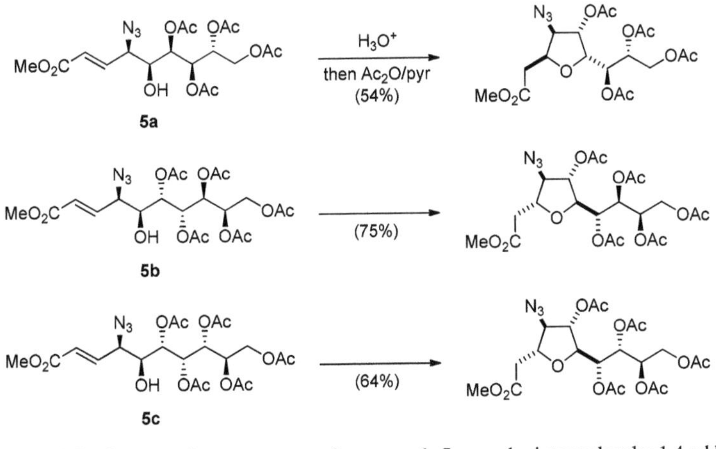

Scheme 56. Acidic cleavage of acetate groups of compounds **5a-c** under intramolecular 1,4-addition.

A typical procedure involved the addition of a small amount of H_2SO_4 (conc.) or HCl 6 M to solutions of compounds **5a-c** in methanol at room temperature over night, followed by neutralization with NaOH 1 M or solid $NaHCO_3$. For simplified purification and NMR spectroscopic analysis, the obtained C-glycosides were subsequently reacetylated. In the case of **5a**, which was used as a mixture of acetyl migration products (see section 2.3.3), an inseparable mixture of products was obtained, presumably composed of 5- and 6-membered rings. In the cases of **5b** and **5c**, only tetrahydofurane derivatives were obtained, contaminated with minor diastereomeric impurities. Since C-glycosides exert no anomeric effect for the stabilization of sterically unfavorable α-anomers, we assumed that the corresponding β-anomers had been formed predominantly (Scheme 56). In order to prove this hypothesis, we tried to prepare crystalline derivatives, which shoul allow the application of x-ray structure analysis. Unfortunately, we were not able to uniformly apply any protecting groups other than acetyl moieties and also the corresponding chloro- and iodo-acetyl protected derivatives (Scheme 57) did not crystallize by applying various techniques such as slow evaporation, vapor diffusion or co-crystallization with (O)PPh$_3$[123].

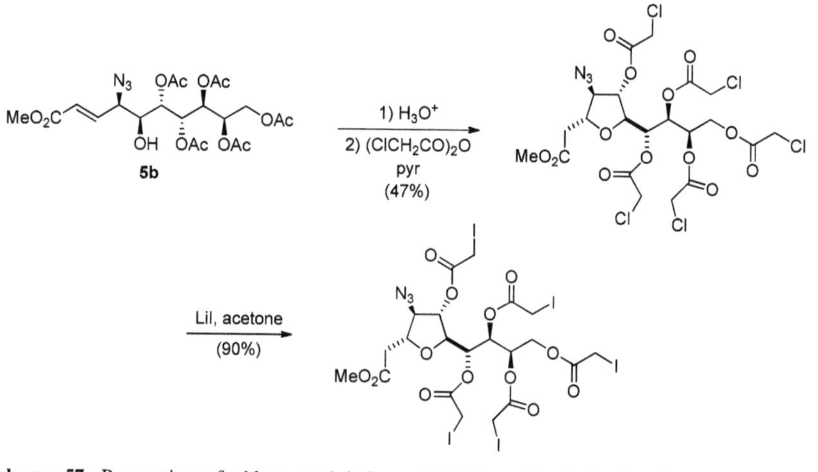

Scheme 57. Preparation of chloro- and iodo-acetyl C-glycoside derivatives for x-ray structure determination.

We then turned to investigate the possibility of performing the acetate cleavage of compounds **5a-c** without triggering the Michael addition as well. Ozonolysis prior to ester cleavage was not an option, since this reaction produced unstable long chained aldehydes, which could be isolated only in the case of **5b** (Scheme 58).

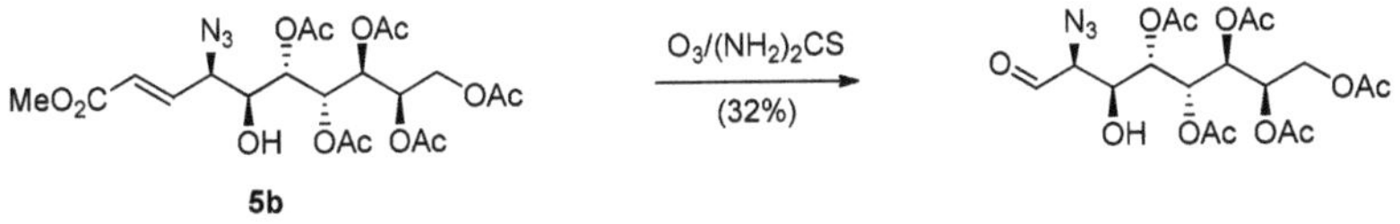

Scheme 58. Ozonolysis of compound **5b**.

Reduction of compounds **5a-c** with complex hydrides also failed, which prompted us to further fine tune the acidic acetate cleavage protocol. Initial attempts showed that the yield of C-glycosides obtained could be enhanced by additionally adding small amounts of water to the reaction mixtures, which led us to the conclusion that the intramolecular 1,4-addition needs a certain amount of water in order to proceed. Subsequently we performed the reaction in methanolic HCl solutions (3 equiv AcCl/MeOH) under anhydrous conditions and after 16 h, TLC analysis showed the emergence of a single new spot. Unfortunately, after neutralization with solid $NaHCO_3$ this new spot was completely converted to the already known C-glycoside spot, which suggested that even the small amounts of water produced during neutralization were sufficient to trigger the Michael addition. Removal of the solvent under reduced pressure without neutralization resulted in decomposition of the product, so we tried to perform the work-up by adding molecular sieves 4 Å to the reaction mixture which is known to also accept HCl. To our delight, this approach was successful and subsequent ozonolysis furnished sugar azides **6a-c** (Scheme 59).

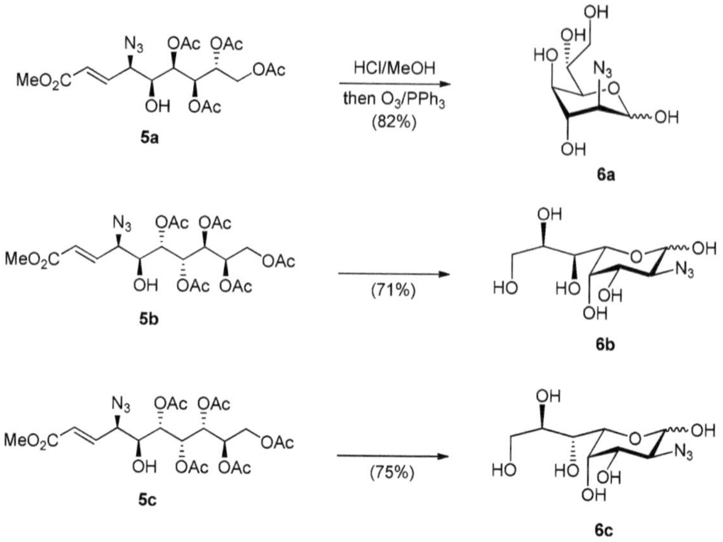

Scheme 59. Deacetylation/ozonolysis of compounds **5a-c**.

The ozonolysis step was performed in MeOH with small amounts of DCM as an indicator. We encountered a solubility problem in the case of the deacetylatation product derived from compound **5b**, which was only soluble in MeOH/water mixtures. Therefore, the ozone stream in this case was bubbled through suspensions in MeOH for a defined amount of time (~ 5 equiv of O_3) and the progress of the reaction was monitored by TLC analysis. If needed, additional ozone was 'added' and finally, the reaction was quenched with PPh_3. Nevertheless, small amounts (~ 3%) of non-ozonolysed product were always isolated in this case. Sugar azide **6a** showed highly complex ^1H and ^{13}C NMR spectra owing to its inherent D-glycero-D-ido-configuration which is known to adopt multiple conformations [124] besides pyranoid and furanoid forms. For example, solution dynamic investigations of pentaacetyl-α-D-ido-pyranose by NMR techniques have shown that this compound has three low energy conformations, namely the $^4C_1$ and $^1C_4$ chair forms and the $^OS_2$ skew-boat form (Figure 9) [124b].

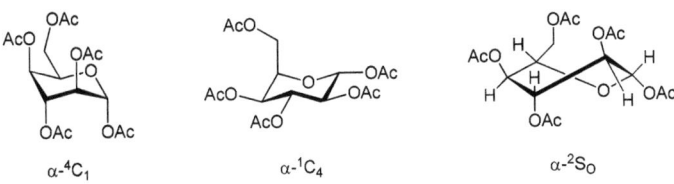

Figure 9. Conformations of pentaacetyl-α-D-ido-pyranose.

Therefore, only anomeric NMR signals were assigned for compound **6a**. Next, we turned to investigate the reduction of the azide moiety of compounds **6a-c**. Unfortunately, DL-dithiothreitol (DTT)/DIPA, thioacetic acid, tributylphosphine/H_2O and H_2/Pd did not (reproducibly) furnish clean products. Owing to the high polarity of the target compounds **8a-c**, purification by standard silica gel chromatography was not feasible. Since we wanted to avoid intricate purification methods like reversed phase HPLC, we chose to reacetylate compounds **6a-c** and subsequently reduced them with DTT/DIPA.

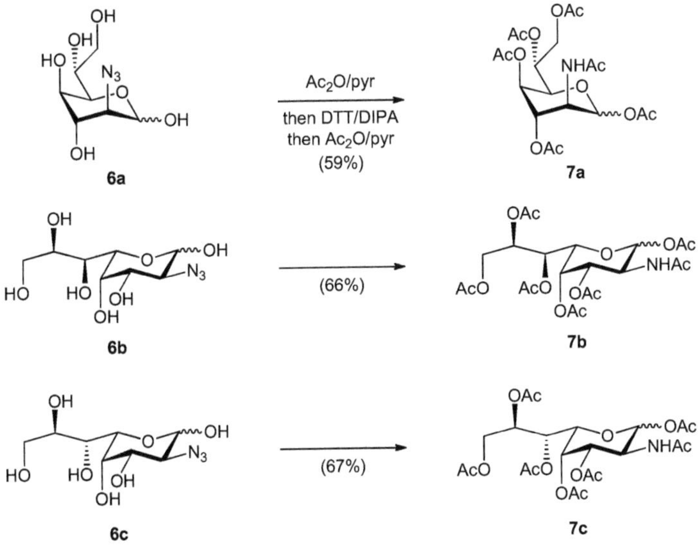

Scheme 60. Azide reduction of compounds **6a-c**.

The azide reduction using DTT has proven to be a comparably fast and reliable method in our hands. The only disadvantage being the required basic conditions, which decompose labile compounds such as **5a-c**. The driving force of the azide reduction with dithiols, besides the evolution of nitrogen is the formation of the corresponding disulfides (Scheme 61).

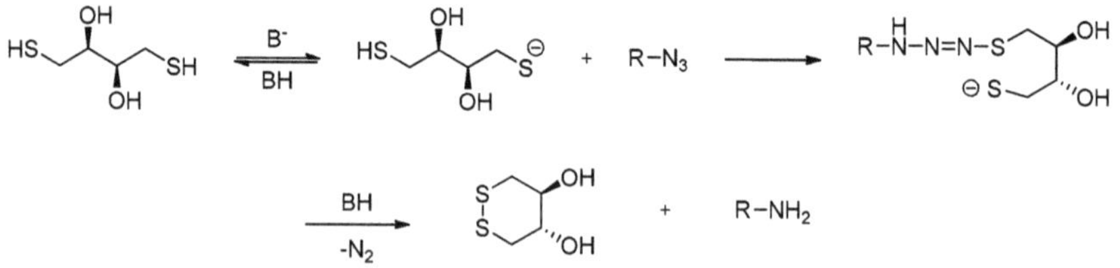

Scheme 61. Reduction of azides with dithiothreitol.

A base is needed in order to deprotonate the dithiol, thus forming a thiolate species which nucleophilically attacks the azide, to produce a triazene intermediate. After de- and reprotonation, the disulfide bridge forms, releasing N_2 and the amine. After acetylation, products **7a-c** could be easily purified by standard column chromatography. In the case of **7a**, two fractions were isolated and exhaustive NMR analysis showed that each of them contained two distinct forms of pentaacetyl-2-acetamido-D-glycero-D-ido-heptose **7a** (Figure 10).

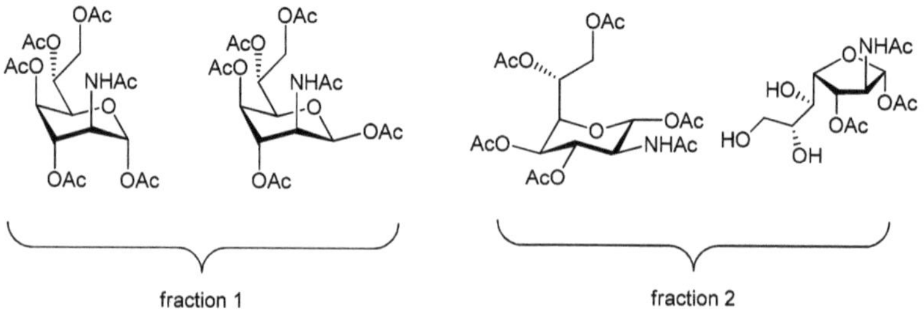

fraction 1   fraction 2

Figure 10. Isolated and characterized isomers of compound **7a**.

Although that **7a** is a known compound[110a], the provided NMR data is scarce since the spectra were recorded on a 90 MHz spectrometer. With all the derivatives available in pyranoid form and their NMR data, we were able to prove the proposed stereochemistry. Since C-4 is incorporated by the starting material, the configurations of C-2 and C-3 can be determined by comparison of the relevant $^3J_{H,H}$ coupling constants (Table 1).

Table 1. Characteristic coupling constants [Hz] of compounds **7a-c**

Compound	$^3J_{1,2}$	$^3J_{2,3}$	$^3J_{3,4}$	$^3J_{4,5}$
7a-α[a]	1.8	3.1	3.1	1.9
7a-β[a]	2.1	2.9	2.9	1.8
7b-α	3.7	11.5	3.3	0.9
7b-β	9.0	11.3	3.5	1.1
7c-α	3.7	11.6	3.3	1.0
7c-β	8.8	11.1	3.3	0.5

[a]: $^4C_1$-pyranoid form.

According to the Karplus relation[125], ^3J coupling constants are smallest at dihedral angles around 90°. In terms of chair-like pyranoid systems, a small coupling constant (0-5 Hz) therefore reflects a *syn* alignment of neighboring hydrogens (axial/equatorial, equatorial/ equatorial), whereas a large coupling constant (7-12 Hz) represents an *anti* alignment (axial/axial). Through comparison of all coupling constants (Table 1) an unambiguous assignment of the stereochemistry is feasible in most cases. However, coupling constants may be in between small and large values (see section 2.4.2), which indicates a distortion of the ideal chair conformation owing to unfavorable steric interactions. This behavior is sometimes encountered when more axial then equatorial substituents are present in a given pyranose, which in some cases leads to ring flip ($^4C_1$ vs. $^1C_4$ conformation). The last step of the structure elucidation is the comparison of calculated (based on the proposed structure) and measured NMR spectra. Additionally, NOE (1,3-diaxial) interactions can be used to provide further proof. With this methodology we could prove the proposed stereochemistry of compounds **6a-b**, **7a-c** and **8a-b**, which additionally was in accordance with the mechanisms of enamine catalysis and palladium π-allyl chemistry used for the construction of the new stereocenters. The last step in our synthesis involved standard cleavage of the acetate protecting groups of compounds **7a-c** with NaOMe in MeOH, furnishing the target compounds **8a-c** in pure form (Scheme 62).

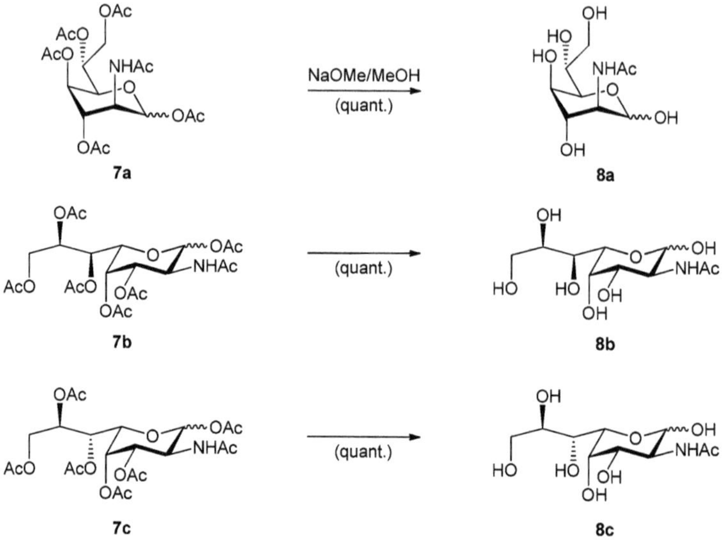

Scheme 62. Zemplén saponification of compounds **7a-c**.

In summary we were able to develop a new approach for the synthesis of 2-amino functionalized heptoses and octoses by subjecting the corresponding pentoses and hexoses to an indium mediated chain elongation. Two new stereocenters were constructed by applying organocatalytic epoxidation, followed by the introduction of nitrogen *via* palladium catalysis. A deprotection sequence was devised subsequently, furnishing either tetrahydrofurane derivatives of the C-glycoside type, or the desired amino sugars by adjusting the reaction conditions of the acidic acetate cleavage step. Thus, the target compounds **8a-c** were obtained in an overall yield of 21-29% over 7 steps.

2.4 Synthesis of fluorinated amino sugars

2.4.1 Epoxidation/fluoride opening approach

We started our investigations by preparing differently O-protected L-serine derived α,β-unsaturated esters **9a-c** (Scheme 63).

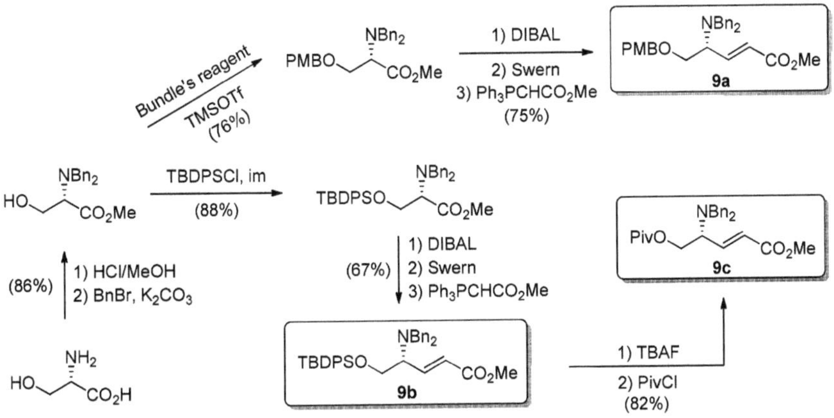

Scheme 63. Preparation of L-serine derived α,β-unsaturated esters.

Since the N,N-dibenzyl protecting group motif was required for the aspired one-pot epoxidation/fluoride opening protocol, we chose to install it first, rather than starting with Garner's aldehyde and switching the protecting groups at a later stage (see section 2.1). It was also known that free hydroxyl moieties lead to partial intramolecular epoxide opening. Thus, we applied different types of O-protecting groups in order to overcome this problem. The PMB group was chosen on account of being orthogonal to N-benzyl; whereas the TBDPS and Piv groups should be reasonable stable under acid conditions. Compounds **9a** and **9b** were prepared according to a literature procedure[126] whereas **9c** was prepared from **9b** by switching the protecting groups, since DIBAL would also reduce the pivaloyl ester. Next, we investigated the hydroxyfluorination of compounds **9a-c**. Unfortunately, it turned out that the mCPBA/HBF$_4$.OEt$_2$

protocol was not suitable for our substrates. In all cases, the O-protecting groups were cleaved off and only cyclized products as described in section 2.1 were isolated. Surprisingly, the TBDPS group was most stable towards HBF$_4$.OEt$_2$, whereas PMB cleavage occurred very readily. Subsequently, we envisioned two alternative strategies; (1) reduction of compound **9a** to furnish the corresponding allylic alcohol, followed by Sharpless epoxidation[127], or (2) substrate controlled epoxidation of **9b** with tBuOOH[128] (Scheme 64).

Scheme 64. Epoxidation of compounds **9a-b**.

The Sharpless epoxidation furnished epoxide **10** as a single diastereomer in good yield. However, the corresponding 'mismatched' epoxide (

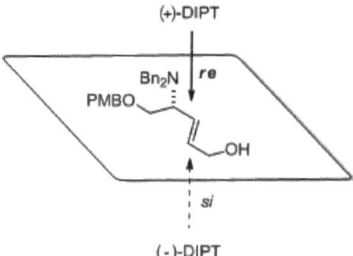

Figure 11) was not accessible by using (+)-diisopropyl tartrate (DIPT), which only resulted in the recovery of starting material.

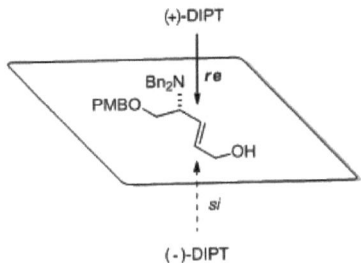

Figure 11. Diastereofacial selectivity of the Sharpless asymmetric epoxidation; dependence on the tartrate enantiomer used.

This observation also suggested that no major epimerization had occurred during the preparation of compounds **9a-c**. Unfortunately, the substrate controlled epoxidation of compound **9b** proved to be highly impracticable. Although the obtained diastereoselectivity was good, yields were below reasonable amounts. Apparently, a major part of the product was lost due to TBDPS deprotection. Owing to the highly basic conditions, the ester moiety of compound **9b** was cleaved in the course of the reaction. Although re-esterification with diazomethane was performed subsequently, only trace amounts of epoxy ester, contaminated with silanol by-product, were isolated and amide **11** was obtained as the major product. However, two different epoxide substrates were then available to be tested in the fluoride opening reaction (Table 2).

Table 2. Attempts at fluoride opening of epoxides **10, 11**.

Entry	Reagent	Substrate	Equivalents	T [°C]	t [h]	Observation
1	$HBF_4 \cdot OEt_2$	10	20	25	1	cyclization
2	$HBF_4 \cdot OEt_2$	10	2	25	1	cyclization
3	$BF_3 \cdot OEt_2$	11	0.33	-20	1.5	no conversion
4	$HBF_4 \cdot OEt_2$	11	0.33	0	1	no conversion
5	$HBF_4 \cdot OEt_2$	11	1	25	5	no conversion
6	$HBF_4 \cdot OEt_2$	11	5	25	144	Partial O-deprotection
7	HF/pyr	11-OH	80	25	5	no conversion

| 8 | TBAF | 11 | 2.5 | 25 | 5 | decomposition |

Unfortunately, all attempts were unsuccessful and again resulted only in cleavage of the O-protecting groups and cyclization. Apparently, the nucleophilicity of the fluoride ion is too low to achieve an epoxide opening under sufficiently mild conditions which leave moderately acid stable protecting groups intact. Although, a successful fluoride epoxide opening without any OH protection had been published[102] previously, we were not able to perform this reaction with our substrates and thus decided to abandon this approach.

2.4.2 Allylation/ozonolysis/α-fluorination approach

The substrate required for Jørgenson's[103] and MacMillan's[104] stereoselective α-fluorination protocols was prepared according to literature procedures[105]. Garner's aldehyde (see section 1.2.3) was allylated with the Roush reagent[129] (Scheme 65), which was freshly prepared by treating trimethyl borate with allylmagnesium bromide followed by (-)-DIPT (see section 3.2).

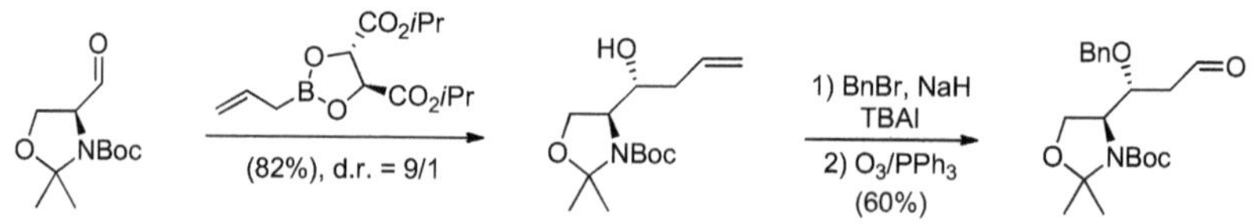

Scheme 65. Roush allylation of Garner's aldehyde, benzylation and ozonolysis.

Owing to the shielding effect of the tartrate ester (Figure 12), the diastereoselectivity of the allylation (reported d.r. > 19/1) is considerably enhanced compared to the corresponding Grignard reaction[130] (d.r. = 3/1).

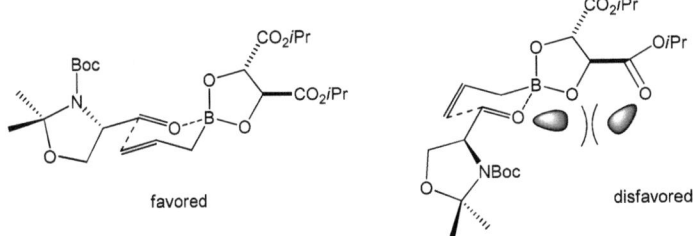

Figure 12. Favored/disfavored transitions states in the Roush allylation; lone-pair repulsion model.

Subsequently, benzylation and ozonolysis were performed, furnishing compound **12**, which was subjected to electrophilic α-fluorination (Scheme 66).

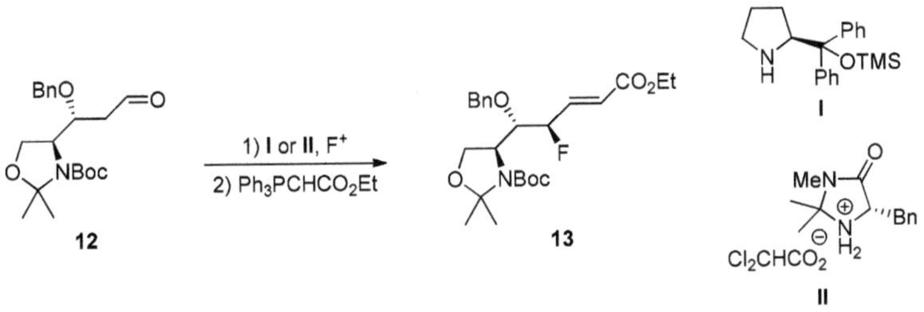

Scheme 66. Organocatalytic α-fluorination of compound **12**.

We chose to directly protect the crude α-fluoro aldehyde by treatment with stabilized Wittig ylide to furnish olefin **13**, in order to avoid potential epimerization. We applied both Jørgenson's proline and MacMillan's phenylalanine derived catalysts in combination with NFSI, which is a cheap, bench-stable electrophilic fluorinating reagent. The proposed catalytic cycle proceeds through the well-known iminium mechanism (Scheme 67, see also section 2.3.2).

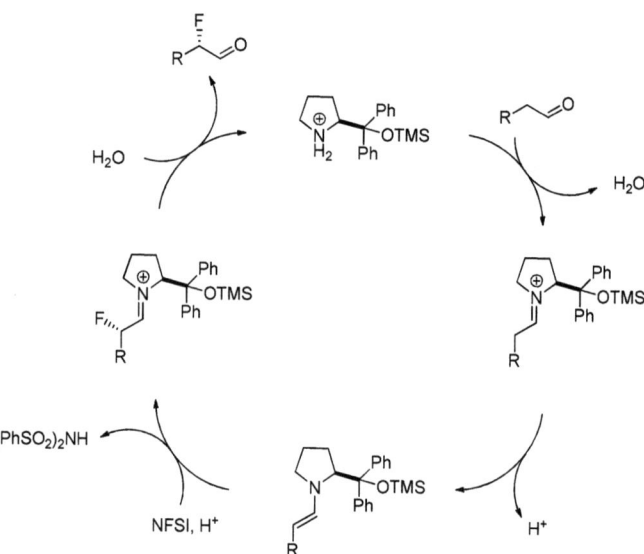

Scheme 67. Electrophilic α-fluorination of aldehydes with chiral amine catalysts.

Although the diastereoselection was excellent (d.r. = 25/1), conversion and overall yields were very low. The best results were achieved with prolinol catalyst **I** (20 mol %) in MTBE at -20°C for 40 h (Table 3).

Table 3. Reagents and conditions for electrophilic α-fluorination.

Entry	Reagent	Catalyst	T [°C]	Yield [%]
1	NFSI	I	-20	20
2	NFSI	I	25	8
3	NFSI	I	60	traces
4	Selectfluor	I	25	-
5	(pyrF)(OTf)	I	25	-
6	NFSI	II	4	-

At -20 °C the reaction was very sluggish, whereas higher temperatures induced decomposition of the product. In all cases TLC analysis of the reaction mixtures showed multiple side products one of which was identified as an elimination product. This observation may indicate that the intermediate α-fluoro aldehydes are very unstable. On the other hand, compound **13** may have partially decomposed upon work-up or purification. An alternative 'work-up' of the reaction by $NaBH_4$ reduction was considered inappropriate regarding the potential selectivity problems of re-oxidation after deprotection. The application of other fluorine sources did not result in productive reactions and also MacMillan imidazolidinone **II** was unsuitable for our substrate. When treated with equimolar amounts of catalyst **II**, aldehyde **12** decomposed readily, whereas the corresponding **I/12** adduct was stable for several hours by ^{1}H-NMR monitoring. We concluded that the success of this approach was hampered by the instability of intermediate open chained α-fluoro aldehydes under the reaction conditions employed. Therefore we reasoned that a *de novo* synthesis of C-2 fluorinated carbohydrates requires a strategy which allows immediate hemiacetal formation of the products.

2.4.3 Stereoselective aldol addition approach[2a]

We started our investigations by preparing fluoroacetyl ephedrine oxazolidinone **14** (Scheme 68) as a chiral auxiliary for the stereoselective aldol addition of serine and threonine derived aldehydes.

Scheme 68. Aldol addition of auxiliary **14** and serine derived aldehydes.

Fluoroacetyl chloride, which is not commercially available from standard suppliers, was prepared by saponification of ethyl fluoroacetate followed by treatment with PCl_5 and distillation. Subsequently, we tested different Lewis acids, bases, and (L-serine derived) amino aldehydes in the aldol reaction (Table 4).

Table 4. Reagents screened for the aldol addition of compound **14**; n.d.: not determined

Entry	Aldehyde	Lewis acid	Base	Yield [%]	d.r.
1	Garner's	TiCl$_4$	TMEDA	68	8/2/1/1
2	Garner's	TiCl$_4$	DIPEA	50	3/2/1
3	Garner's	nBu$_2$BOTf	DIPEA	15	n.d.
4	Garner's	nBu$_2$BOTf	TEA	-	-
5	Garner's	-	LDA	18	n.d.
6	Garner's	-	LiOH	-	-
7	Garner's	Sn(OTf)$_2$	LDA	-	-
8	Garner's	nBu$_3$SnBr	LDA	n.d.	n.d.
9	N,N-Bn$_2$-O-TBDPS	TiCl$_4$	DIPEA	-	-
10	N,N-Bn$_2$-O-TBDPS	nBu$_2$BOTf	TEA	-	-
11	N,N-Bn$_2$-O-TBDPS	nBu$_3$SnBr	LDA	-	-

The N,N-dibenzyl protected aldehyde was completely decomposed in all cases and no aldol products were isolated. Delightfully for us, Garner's aldehyde in combination with most Lewis acids afforded the desired fluorohydrin product (Scheme 68). Lithium and boron reagents furnished only very low yields and also the selectivity was presumably low. Tin(II)triflate proved to be unreactive and resulted only in the recovery of starting material, whereas tin(IV) furnished minor amounts of product, although the residual alkyl tin species could not be removed subsequently. The best results were achieved with TiCl$_4$ (Table 4, entries 1, 2). As observed by other groups[70], when DIPEA was used as a base, the obtained diastereoselectivity was only moderate, whereas TMEDA furnished an 8/2/1/1 mixture of the four possible diastereomers (Figure 13). An excess of auxiliary (1.3 equiv), TiCl$_4$ (1.4 equiv) and base (4 equiv) was used, nevertheless a considerable amount of starting aldehyde was recovered in all cases. We found that the optimal temperature range for the aldol addition was between -40 and -20 °C. Therefore, the viscid brown-black reaction mixtures were left to slowly warm from -40 to 0 °C and finally subjected to aqueous work-up and purification by flash-column chromatography to furnish compound **15** as a colorless to light yellow crystalline solid with minor diastereomeric impurities.

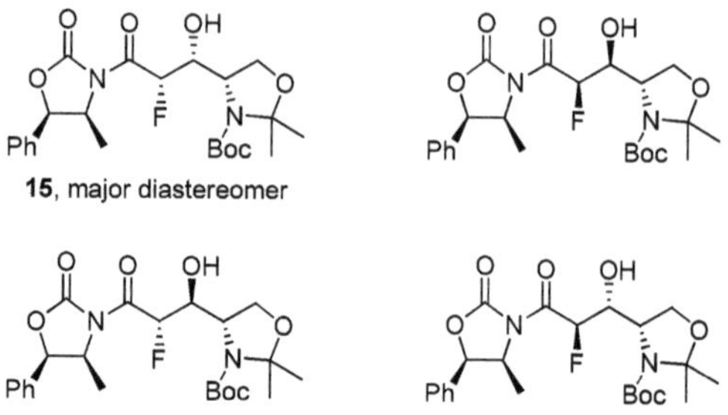

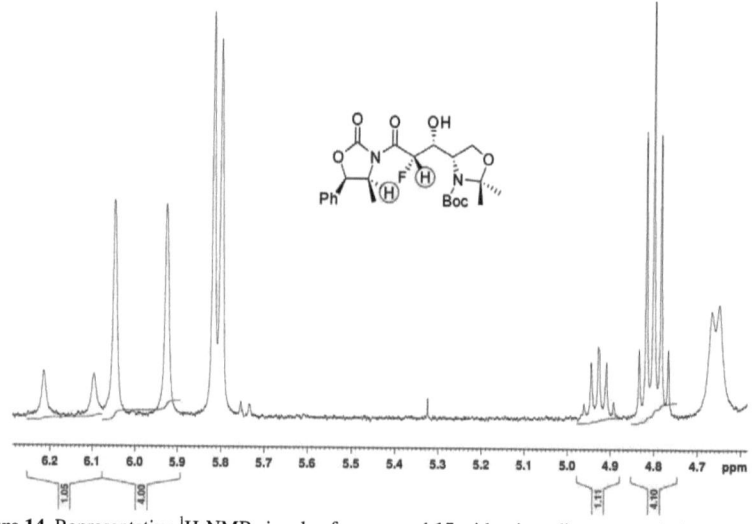

Figure 13. Possible diastereomers resulting from the aldol addition of **14** and L-Garner's aldehyde.

The diastereomeric ratios were estimated based on the amount of obtained material after column chromatographic separation. Since compound **15** could not be obtained in diastereomerically pure form, the relative ratios were determined by integration of representative ^{1}H-NMR signals (Figure 14).

Figure 14. Representative ^{1}H-NMR signals of compound **15** with minor diastereomeric impurity.

Unfortunately, neither ^{1}H nor ^{19}F-NMR spectra of crude products could be used to determine the diastereomeric ratio, owing to signal overlap. We did not prove the stereochemistry of compound **15** at this stage of the synthesis since we envisioned an argumentation based on the final pyranoid products as described for compounds **7a-c** in section 2.3.4. Titanium behaves as a non-chelating metal (see section 1.2.2) in this reaction, furnishing Evans-*syn* aldols, owing to the formation of the kinetic Z-enolates. Assuming that the auxiliary overrides the facial selectivity of the aldehyde, this example nevertheless represents a mismatched double asymmetric induction scenario. However, we were confident, that the selectivity would be enhanced by applying the corresponding matched D-amino acid derived aldehydes. Thus, D-serine as well as D-, L- and D-*allo*-threonine derived aldehydes were subsequently prepared and subjected to our titanium mediated aldolization (Scheme 69).

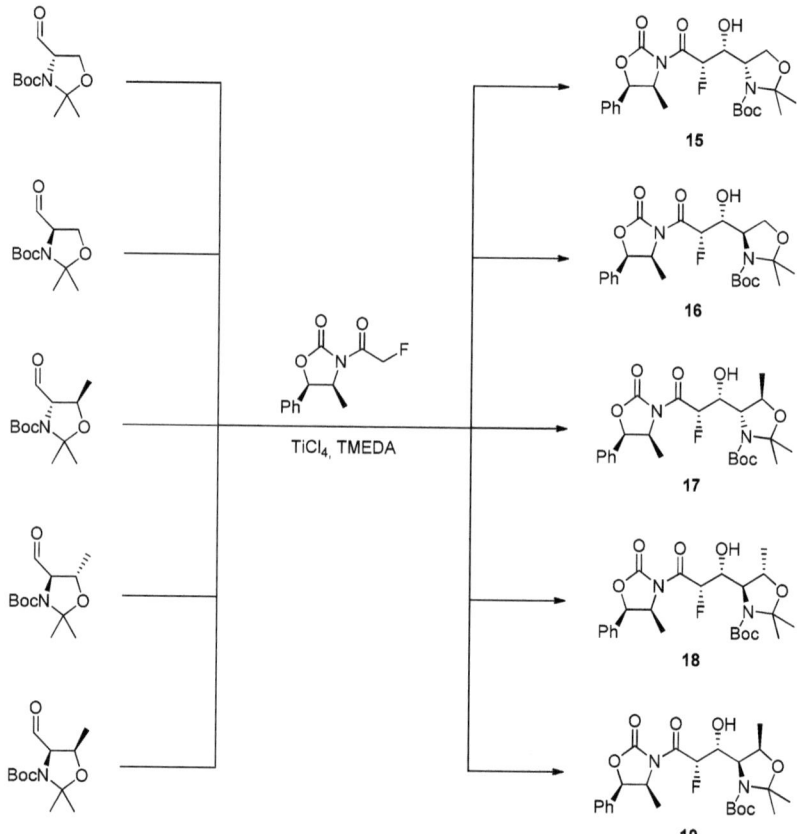

Scheme 69. Preparation of fluorohydrins **15-19**.[2a]

Delightfully for us, D-amino acid derivatives displayed high diastereoselectivity (Table 5), thus forming two stereocenters and the fully functionalized carbon backbone in a single step.

Table 5. Yields and selectivities for fluorohydrins 15-19.

Compound	Isolated yield [%]	Combined yield of diastereomers [%]	Total yield brsm [%]	d.r.
15	45[a]	68	83	8/2/1/1
16	56	66	82	17/1/1/1
17	33	44	77	15/3/1/1
18	50	55	76	32/2/1/0.5
19	45	47	47	20/1

[a]: Calculated value, compound 15 could not be completely separated from one minor diastereomer, which was removed at a later stage of the synthesis; brsm: based on recovered starting material

Compounds 15 and 16 were obtained in higher yields than the corresponding threonine derivatives 17, 18 and 19 although these in turn provided higher diastereomeric excess. The best results in terms of selectivity were achieved with D-*allo*-threonine derivative 19, which furnished only two diastereomers in a ratio of 20/1, although in this case no starting material was left. In general, only low amounts (3-5%) of product were lost due to Boc deprotection in the course of the reaction. Since the aldol products obtained feature acid labile protecting groups, we subsequently investigated their cleavage with aqueous or methanolic HCl. Unfortunately the liberated amine moiety displaced the oxazolidinone auxiliary under these conditions, furnishing a γ-lactam product (Scheme 70).

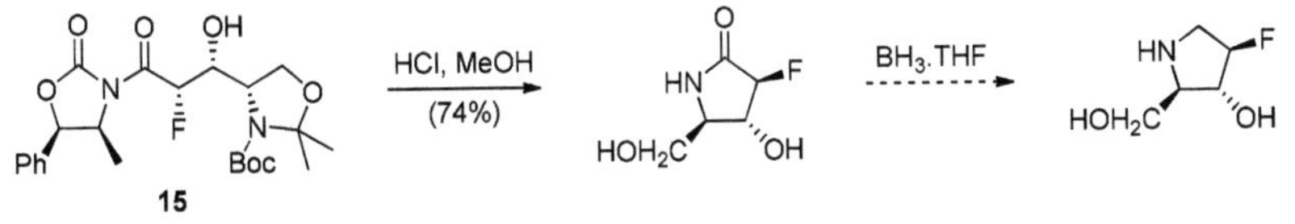

Scheme 70. Deprotection of compound 15 with methanolic HCl; formation of γ-lactam species.

Although additional epimerization (d.r. = 2/1) was encountered in HCl/MeOH, the resulting γ-lactam product represents an interesting synthetic target, since subsequent reduction with $BH_3.THF$ [131] would furnish fluorinated imino sugars[132], which comprise potent pentosyl-transferase inhibitors and may be used for example in the treatment of mycobacteria induced diseases such as tuberculosis. Nevertheless, we focused on devising a deprotection protocol, which should furnish the desired 4-amino-2-fluoro-pentoses and hexoses. We reasoned that a selective cleavage of the isopropylidene beside the Boc moiety was required in order to overcome the problem of lactam formation. The use of acetic acid (80%) at elevated temperatures in this respect gave inconsistent results, whereas acidic ion exchange resin reproducibly cleaved the acetonide protecting group in a spot to spot reaction leaving the Boc group intact. Unfortunately, we were not able to perform this reaction with compound **19**, which resulted only in the recovery of starting material. Subsequently, the oxazolidinone auxiliary was substituted with NaOMe in MeOH at -40 to -25 °C furnishing esters **20-23**. For convenience, the order of these two steps can be switched, which results in similar yields (Scheme 71).

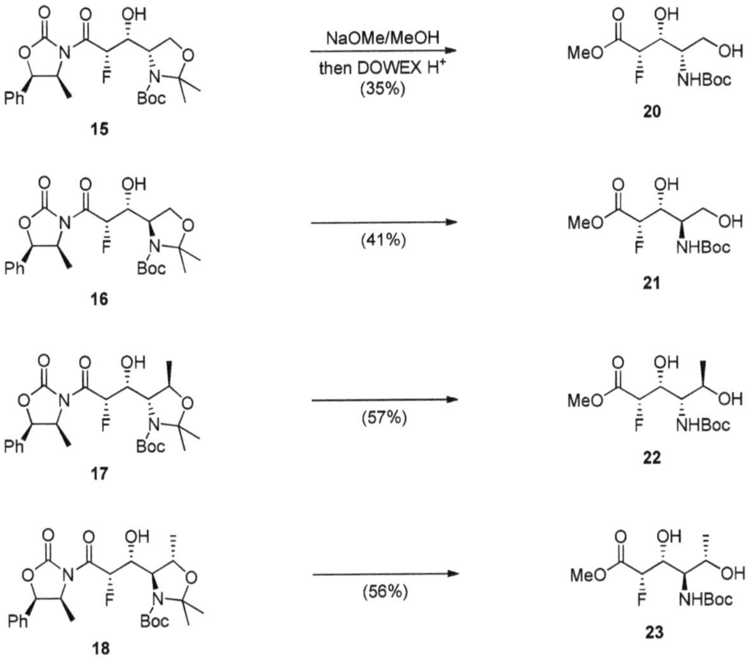

Scheme 71. Deprotection of fluorohydrins **15-18**; oxazolidinone and isopropylidene cleavage.

For reasons, not fully comprehensible to us the overall yield of oxazolidinone and isopropylidene cleavage was worse than the yield of the individual steps ($\geq$ 90% each) combined, independent of their order of execution. We suggest two possible explanations for this behavior. (1) Ester products **20-23** were partially lost upon column chromatographic purification owing to their relatively high polarity. (2) The ion exchange resin induced partial cleavage of the Boc group, retaining the free amine products, which were therefore not visible upon TLC analysis. There are two indicators which render explanation (1) most plausible. On the one hand, more apolar threonine derivatives **22** and **23** were obtained in higher yields and on the other hand, acetonide cleavage of compound **19** after very long reaction times resulted in the formation of small amounts of product upon TLC analysis, which were not found after column chromatographic purification even when rinsed with MeOH. Explanation (2) seems additionally unlikely since DOWEX H⁺ deprotection of compounds **15-18** resulted in nearly quantitative yields. Nevertheless, with esters **20-23** in hands we were then able to perform DIBAL reduction, followed by acetylation and Boc cleavage to furnish pyranoid compounds **24-27** (Scheme 72).

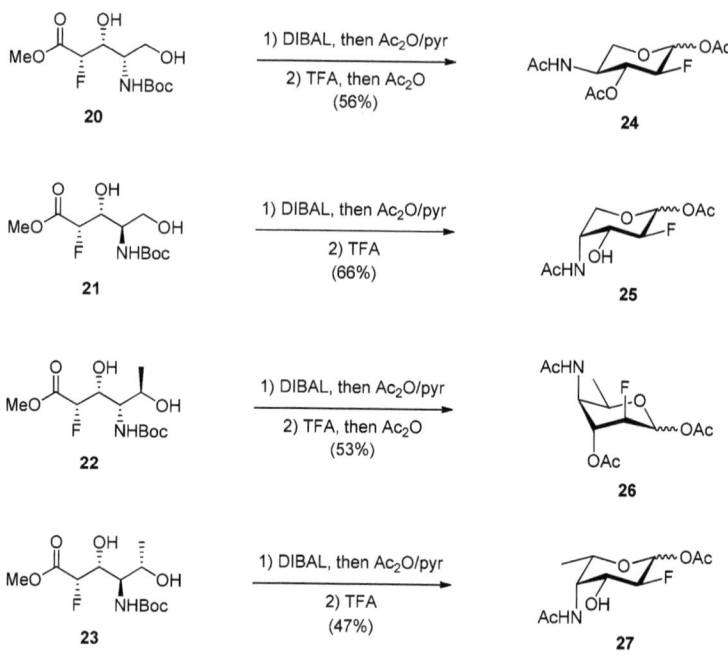

Scheme 72. DIBAL reduction and Boc cleavage of compounds **20-23**; preparation of acetylated sugars **24-27**.

For a successful reduction it proved to be crucial to use fresh DIBAL reagent in superstoichiometric amounts (4-6 equiv). Nevertheless, the conversion of esters **20-23** was not always complete and partial over-reduction to the corresponding alcohols occurred. Subsequently, acetylation was performerd in order to trap the compounds in their pyranoid form, since acidic Boc deprotection of the free carbohydrates would result in imine formation or decomposition owing to the 2-fluoro aldehydes being present in equilibrium. Compounds **24-27** were then treated with TFA followed by neutralization with basic ion exchange resin. In the case of D-amino acid derivatives acetate migration (C-3-OAc → C-4-NH₂) readily occurred to furnish anomeric acetates **25** and **27**. L-amino acid derived compounds on the other hand were treated with Ac$_2$O after Boc cleavage to furnish fully acetylated compounds **24** and **26**, since the migration did not occur in these cases. This behavior can be rationalized considering the inherent *syn* relationship between the groups involved in compounds **25** and **27** *versus* the *anti* alignment within compounds **24** and **26**. Interestingly, in these 'non-migration' cases slow autocatalytic cleavage of C-3-OAc instead occurred. Finally, Zemplén saponification afforded the target compounds **28-31** (Scheme 73).

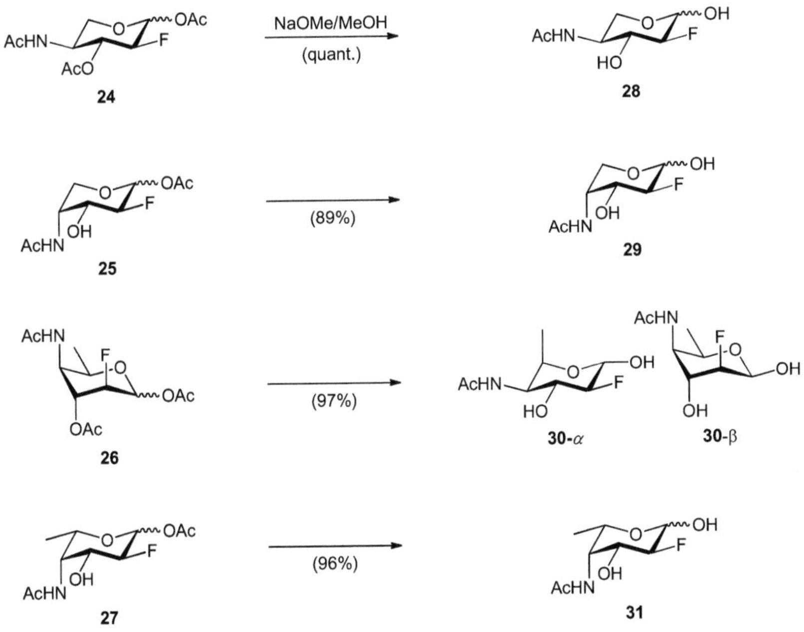

Scheme 73. Zemplén saponification of compounds **24-27**; synthesis of fluorinated carbohydrates **28-31**.

The configurations of compounds **28-31** were proven by comparison of their characteristic $^3J_{H,H}$ coupling constants (Table 6) in a similar fashion as described in section 2.3.4.

Table 6. Characteristic coupling constants [Hz] of compounds **27-30**.[2a]

Compound	$^3J_{1,2}$	$^3J_{2,3}$	$^3J_{3,4}$	$^3J_{4,5}$
28-α	3.7	9.0	9.9	6.2, 10.1
28-β	7.8	8.8	10.0	5.2, 10.5
29-α	3.1	8.4	4.4	3.0, 4.5
29-β	7.2	9.2	4.9	2.2, 2.5
30-α	5.0	6.4	7.3	4.4
30-β	1.0	3.4	3.0	2.4
31-α	4.1	10.3	4.7	1.7
31-β	7.8	9.8	4.9	1.6

Interestingly, D-ido configurated compound **30**-α seems to adopt a (distorted) $^1C_4$ chair-like conformation, whereas the corresponding β-anomer features the $^4C_1$ chair. However, both anomers of acetylated compound **26** adopt the $^4C_1$ form.

In summary, we developed a new approach for the synthesis of fluorinated amino sugars by applying a titanium mediated aldol addition on serine and threonine derived aldehydes with fluoroacetyl ephedrine oxazolidinone. After sequential acidic cleavage of the protecting groups, the target compounds were obtained in an overall yield of 16-23% over seven steps. In particular, carbohydrates **29** and **31**, arabino- respectively galacto-configurated products represent interesting compounds since their parent amino sugars (4-amino arabinose, tomosamine, see section 2.2.2) are naturally abundant.

2.4.4 Further elaboration on the aldol addition approach

After the successful preparation of four different serine and threonine derived fluorinated carbohydrates we wanted to further expand our methodology by additionally applying the corresponding cysteine derived aldehydes[2b], which were prepared according to a literature procedure[133] and subsequently subjected to our titanium mediated aldol reaction (Scheme 74).

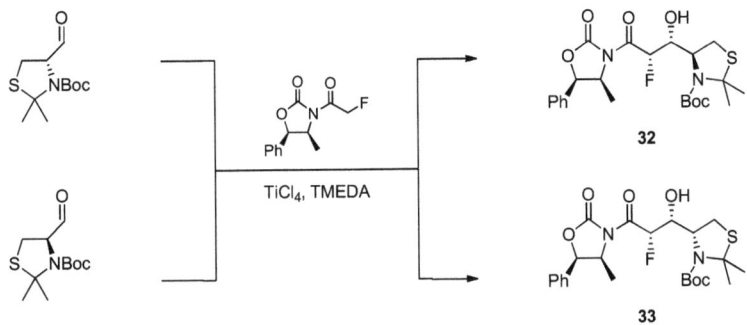

Scheme 74. Aldol addition of cysteine derived aldehydes.[2b]

Obtained yields were similar compared to the corresponding serines but to our complete surprise, the asymmetric induction of the cysteine aldehydes was inverted (Table 7).

Table 7. Yields and selectivities for cysteine derived fluorohydrins **32** and **33**.

Compound	Isolated yield [%]	Combined yield of diastereomers [%]	Total yield brsm [%]	d.r.
32	49	65	89	5/1.2/0.2/0.2
33	59	68	90	14/1/0.7/0.4

In this instance, L-cysteine apparently represents the matched case, whereas L-serine constitutes a mismatched case scenario (see section 2.4.3), which can be deduced by comparison of the d.r. values of the respective substances. Owing to the superior stability of the Ti-S bond, which was also harnessed by Crimmins *et al*[70] in their aldol additions with oxazolidine-thiones (see section 1.2.2), a chelating transition-state seems to be involved in reactions of cysteine derived aldehydes, which prefers the formation of Chelate-Cram, rather than Felkin-Ahn products. Thus, we propose a $^4C_1$ chair-like transition-state with the thiazolidin moiety in an axial orientation, allowing the sulfur to coordinate to titanium (Figure 15). Owing to this chelate effect, we reasoned that our proposed TS would be favored, despite adverse steric interactions. When a similar TS is assumed for the corresponding mismatched D-cystein derivative, N,O-lone-pair repulsion might account for the overall lower selectivity in this case (Figure 15).

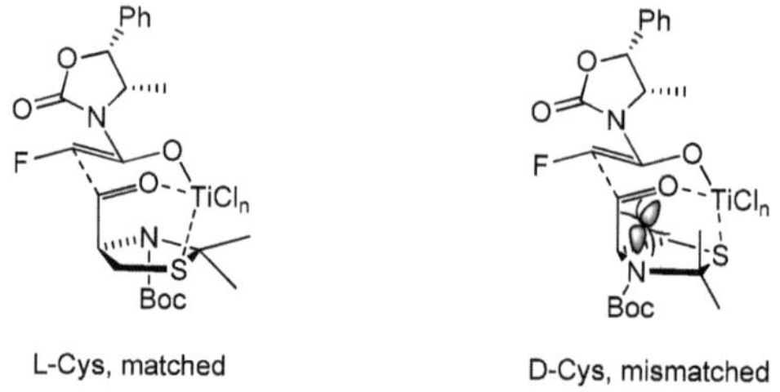

L-Cys, matched   D-Cys, mismatched

Figure 15. Proposed transition-states for aldol additions with L- and D-cystein derivatives.[2b]

A comparable behavior was also encountered in Kobayashi modified Mukaiyama aldol reactions (see section 1.2.1). 1,3-dithiane substituted aldehydes in this case led to the formation of *syn* instead of *anti* products with inverted diastereofacial selectivity[54], which indicates the high propensity of titanium to form chelating transition-states with sulfur containing substrates.

Subsequently, we turned to investigate the deprotection of compounds **32** and **33**. The possibility to prepare thiosugars from these compounds was impeded by the fact that the isopropylidene protecting group in these cases was more stable than the Boc group. The application of DOWEX H⁺ resin resulted only in the recovery of starting material, whereas other reagents such as HCl 3 M furnished mixtures of partially deprotected products (Scheme 76). Considering the problem of lactam formation (see section 2.4.3), we devised a different strategy, based on the sulfur present in compounds **32** and **33**. A Pummerer-type rearrangement (Scheme 75) should be performed, to finally furnish 2-amino-4-fluoro-pentoses.

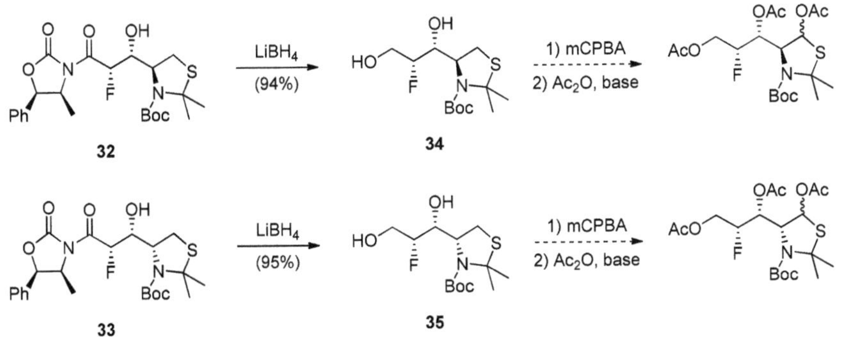

Scheme 75. Reductive cleavage of auxiliaries and initial attempts at Pummerer rearrangement.

Thus, reductive cleavage of the auxiliaries was performed, furnishing compounds **34** and **35**, which were subsequently oxidized and subjected to rearrangement. Unfortunately, we were not able to accomplish this reaction with Ac$_2$O under basic conditions. Additionally, the use of mCPBA resulted in partial over-oxidation, generating sulfone byproducts. We observed that Ac$_2$O in the presence of NaOAc or pyr resulted only in OH acetylation. After prolonged reaction times at elevated temperatures, traces of a product bearing four acetate moieties could be isolated. Although the structure of this compound remained unclear, it was evident that the isopropylidene group had been lost. Thus, we reasoned that an open-chained compound was required in order to achieve the desired rearrangement. Therefore, compound **35** was treated with HCl 3 M and subsequently re-protected using Sanger's reagent (Scheme 76).

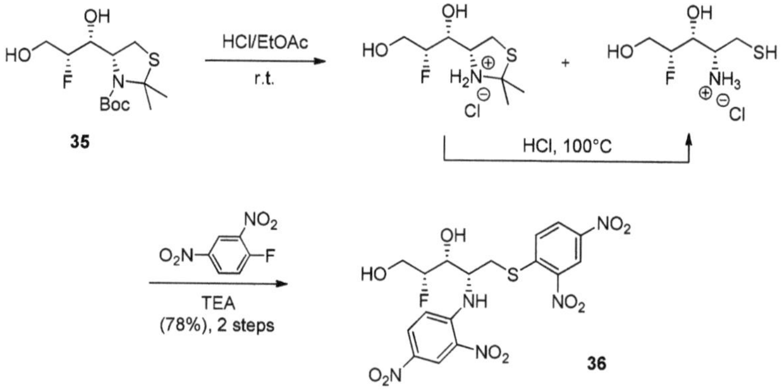

Scheme 76. Acidic deprotection of compound **35** and treatment with 1-fluoro-2,4-dinitrobenzene.

The acidic deprotection did not proceed completly at room temperature and afforded a mixture of products in ratio of ~ 60/40. Only when heated to 100°C, remaining acetonide protected compound could be converted to the free sugar alcohol, which was subsequently treated with two equiv of 1-fluoro-2,4-dinitrobenzene (Sanger's reagent) to furnish compound **36** as a bright yellow crystalline solid. The OH groups in this case remained unprotected since their nucleophilicity is not sufficiently high to perform the aromatic substitution and the N-dinitrophenyl protecting group is potentially cleavable with basic ion exchange resin. When only one equiv of Sanger's reagent was used, the N-protected free thiol was obtained predominantly, suggesting that a selective S-protection might be elusive. After oxidation with mCPBA (Scheme 77), various reagents and conditions for the aspired Pummerer rearrangement were screened (Table 8). Unfortunately, no desired products could be obtained.

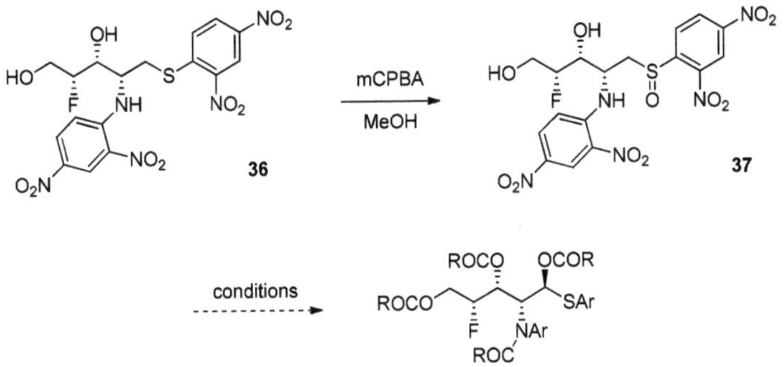

Scheme 77. Oxidation of compound **36** and attempted rearrangement of **37**.

The mCPBA oxidation of compound **36** was performed in MeOH, since reaction product **37** crystallized from MeOH and no over-oxidation was observed. The reaction was typically complete after 2 h at 0 °C and subsequently subjected to basic aqueous work-up. However, the progress of the reaction could not be monitored by TLC, since compound **37** interestingly had the same Rf-value as starting material **36**. Crude **37**, was subsequently treated with different bases and acetic- or trifluoroacetic anhydride (Table 8).

Table 8. Reagents and conditions for the Pummerer rearrangement of compound **36**.

Entry	Base	Acid anhydride	Solvent	T [°C]	Observation
1	NaOAc	Ac$_2$O	Ac$_2$O	130	decomposition
2	pyr	Ac$_2$O	Ac$_2$O/pyr	25	decomposition
3	-	Ac$_2$O	THF/Ac$_2$O	25	mixture of mono- and diacetylated products
4	-	Ac$_2$O	THF	25	low conversion
5	pyr	Ac$_2$O	THF	25	mixture of mono- and diacetylated products
6	2,4-lutidine	Ac$_2$O	MeCN	0-25	mixture of mono- and diacetylated products
7	TEA	TFAA	MeCN	0	decomposition
8	2,4-lutidine	TFAA	MeCN	0-25	mixture of mono-, diacetylated products and epimerization
9	2,4,6-collidin	TFAA	MeCN	0-25	mixture of mono-, diacetylated products and epimerization

Unfortunately, we found that the dinitrophenyl protecting group was unsuitable for our purposes, since we were not able to establish conditions which induced the rearrangement of **37** but did not cleave the protecting groups, or cause epimerization to furnish multiple diastereomers.

Thus, we turned our attention on a singlet oxygen mediated rearrangement[134] of compounds **34** and **35**. As observed by other groups, this photochemical Pummerer reaction seems to be superior for the oxidation of thiazolidine derivatives compared to the conventional conditions. Treatment with Ac$_2$O or silyl triflates for example led to unexpected ring expansions [135] or partial eliminations[136] in some cases (Scheme 78).

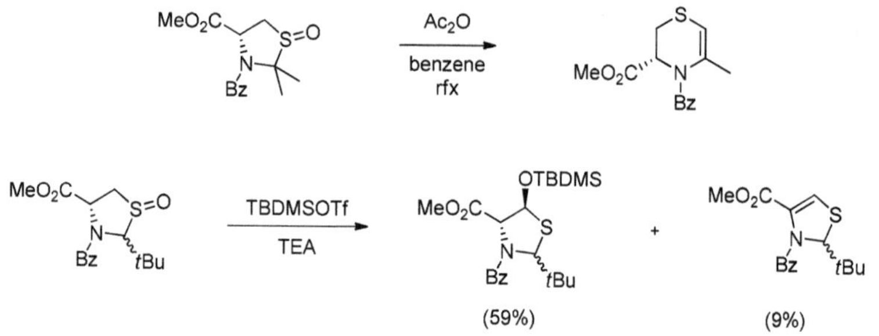

Scheme 78. Conventional Pummerer rearrangement of thiazolidine S-oxides; ring expansion and elimination.

However, the more elegant photochemical approach, first described by Ando *et al*[137] in 1984 furnishes the desired rearrangement products in high yields and only low amounts of sulfoxide side products are formed under optimized conditions (Scheme 79).

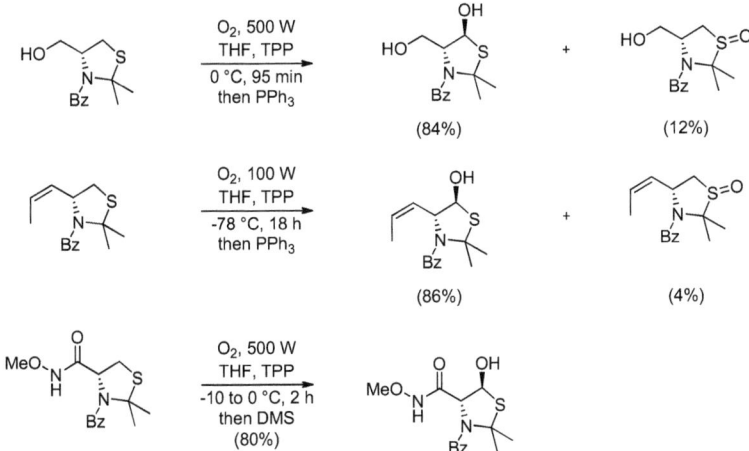

Scheme 79. Representative examples for the photochemical Pummerer-type rearrangement of thiazolidines.

The reaction involves the generation of singlet oxygen *via* a photosensitizer and visible light irradiation. The dye of choice used in most cases is 5,10,15,20-Tetraphenyl-21*H*,23*H*-porphine (TPP), although methylene blue and polymer supported Rose bengal have also been successfully applied. These photosensitizers act by absorbing light in the visible range and subsequently transferring the energy absorbed to oxygen, thus inducing spin inversion ('triplet-triplet annihilation'). Since most organic molecules adopt a singlet electronic ground state, they are more reactive towards singlet oxygen according to the selection rules[138]. The proposed reaction mechanism proceeds through a sulfoperoxide species. After proton shift and rearrangement a hydroperoxide intermediate is formed, which is stable at 0 °C (Scheme 80). This intermediate is finally reduced with PPh₃ or DMS to yield the monothioacetal products.

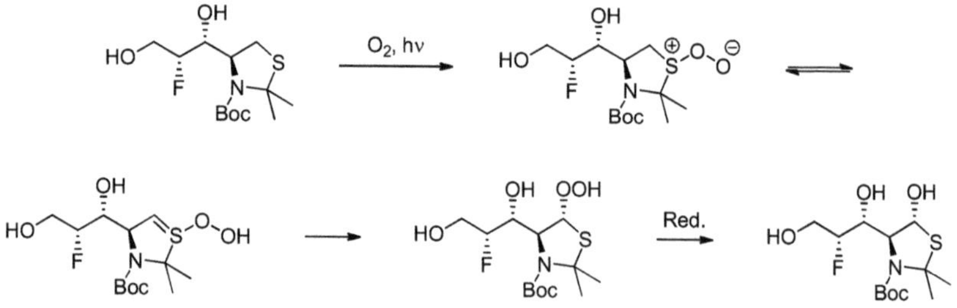

Scheme 80. Mechanism for photochemical Pummerer-type rearrangement.

Besides amino acids and carbohydrates, this methodology also provides access to β-lactam derivatives such as penicillins[134b] (Scheme 81).

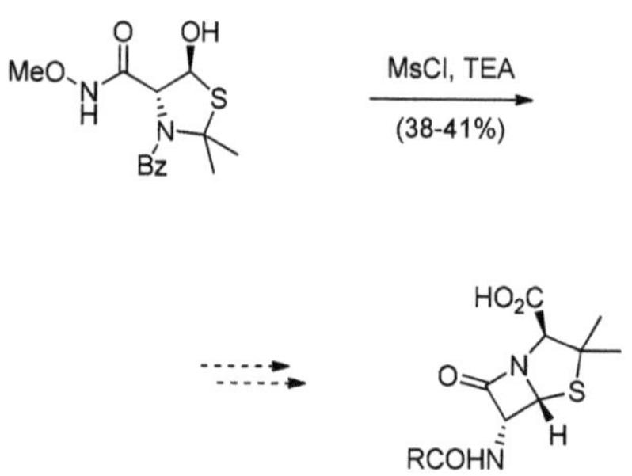

Scheme 81. Synthesis of β-lactams from cysteine derived, photo-Pummerer functionalized compounds.

The product distribution and yield of the photo-Pummerer reaction is also highly dependent on the nature of the solvent. When protic solvents such as MeOH are used for example, sulfoxide products are predominantly formed. Depending on the substrate, useful solvents for this reaction found include benzene, toluene, THF and MTBE. In our hands, toluene and THF proved to be inferior compared to MTBE, which is additionally less hazardous concerning the formation of ether peroxides. As a light source we used a 500 or 100 W halogen lamp and covering with aluminum foil ensured optimal irradiation. The oxygen stream was supplied by an oxygen concentrator device (range: 80-93% O_2). The temperature did not seem to have a significant impact on the reaction as long as it was kept below 0 °C. Thus, the reactions were conveniently performed at -78

°C. However, the irradiation time and the power of the lamp proved to be crucial (Table 9).

Table 9. Photochemical oxidation; dependence of yield on reaction time and power of halogen lamp.[2b]

Entry	P [W]	t [h]	yield [%]	yield brsm [%]
1	500	4	23	89
2	500	2	22	88
3	500	1	12	98
4	100	8	15	84
5	100	4	12	87

brsm: based on recovered starting material

Unfortunately, the conversion of our substrates was very low, compared to literature procedures. In our case, the reaction did not proceed any further after 2-4 h, resulting in the recovery of large amounts of starting material. However, almost no material was lost in this manner and after several reaction cycles acceptable amounts of product were obtained (Table 9, entry 3, five cycles: 51%, 89% brsm; entry 2, two cycles: 41%, 77% brsm). We observed a characteristic color change from wine red to green in the course of the reaction and the original red coloring was restored after reductive work-up. Column chromatographic separation of product and starting material was subsequently performed, furnishing the desired compounds **38** and **39** as single diastereomers (Scheme 82), contaminated with traces of dye.

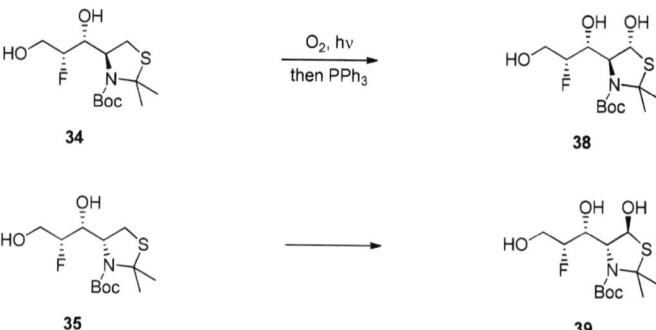

Scheme 82. Photochemical Pummerer-type rearrangement of compounds **34** and **35**.

The stereochemistry of the newly formed stereocenter was not proven since it was degraded after cleavage of the acetonide moiety. In general, all sulfoxides presented in this section were obtained as single diastereomers, suggesting that the amine moiety effectively shields one face of the sulfur from an attack of the oxidant. Finally, compounds **40** and **41** were treated with HCl as described above and subsequently acetylated to furnish the target 2-acetamido-4-fluoro pentoses (Scheme 83).

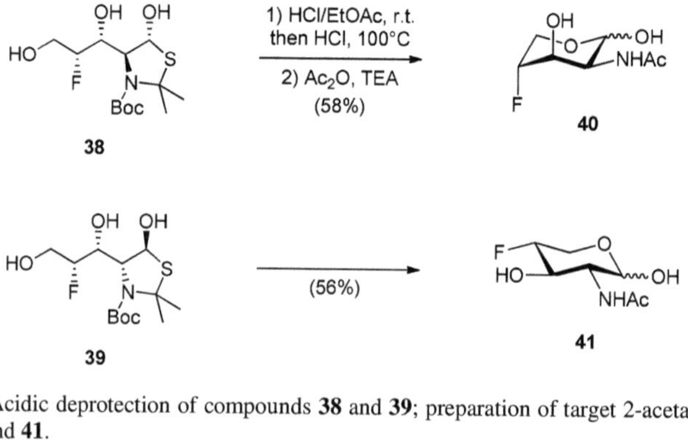

Scheme 83. Acidic deprotection of compounds **38** and **39**; preparation of target 2-acetamido-4-fluoro pentoses **40** and **41**.

The configurations of compounds **40** and **41** were again proven by analysis of the coupling constant pattern (see sections 2.3.4 and 2.4.3) (Table 10).

Table 10. Characteristic coupling constants [Hz] of compounds **39** and **40**.[2b]

Comp.	$^3J_{1,2}$	$^3J_{2,3}$	$^3J_{3,4}$	$^3J_{4,5a}$	$^3J_{4,5b}$	$^3J_{4-F,5a}$	$^3J_{4-F,5b}$	$^3J_{4-F,1}$
40-α	7.7	3.4	4.8	1.9	3.1	32.9	14.9	1.5
40-β	3.0	4.0	6.0	5.0	2.9	11.1	26.7	-
41-α	3.3	9.9	8.0	6.1	9.3	n.d.	n.d.	3.3
41-β	8.0	10.2	8.3	9.9	5.5	4.4	2.7	-

H-5a and H-5b: assigned by order of shift; n.d. not determined.

Interestingly, D-lyxo configured compound **40** adopts a (distorted) $^1C_4$ chair-like conformation, despite featuring two, respectively three axial and one (two) equatorial substituents. We propose that the axial fluorine moiety in this case causes a stabilizing hyperconjugative interaction with the axial hydrogen at C-5. Therefore, the *gauche* conformation is preferred over the corresponding *anti* alignment (Figure 16), which is for example also observed in 1,2-difluoroethane[139].

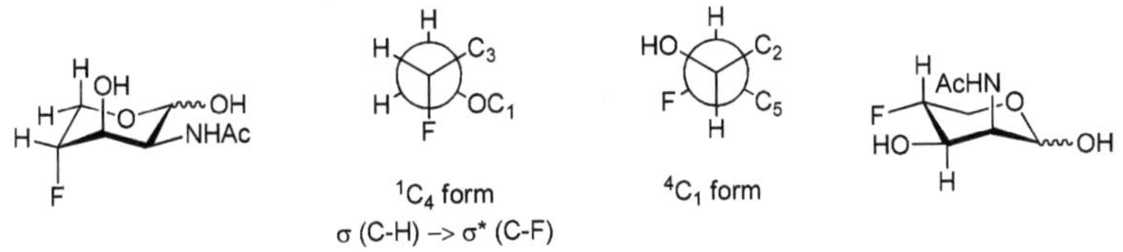

Figure 16. Gauche ($^1C_4$) and anti ($^4C_1$) conformations of compound **40**.

D-xylo configured compound **41** however does not adopt the *gauche* ($^1C_4$) conformation. In this case the adverse steric effects of the axial substituents seem to outweigh the hyperconjugative stabilization. This effect can also be seen on the relatively large H5-F coupling constants of compound **40**. However, a similar behavior for 2-fluoro sugars (section 2.4.3) was not observed, since the hyperconjugation of the ring oxygen (anomeric effect) presumably outweighs the H1-F interaction in this case (Figure 17).

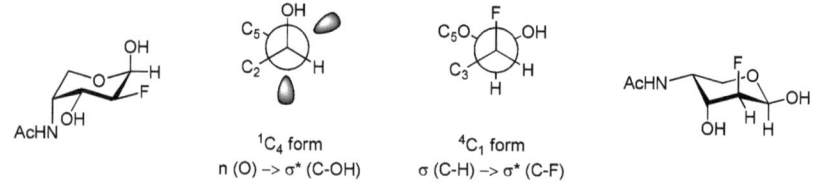

Figure 17. Conformeres of compound **29-α**; anomeric vs. fluorine-*gauche* effect.

In summary, we were able to further expand our titanium mediated aldol reaction methodology on cysteine derived aldehydes for the preparation of 4-acetamido-2-fluoro-pentoses by applying a photochemical Pummerer-type rearrangement. The target compounds were obtained in an overall yield of 47-48% over four steps. Although the conversion of the rearrangement step was low, almost no material was lost when short irradiation times were applied. Thus, after multiple reaction cycles acceptable yields resulted. Additionally, we observed an unexpected change of selectivity in aldol additions with cysteine derivatives, which led to the preferential formation of Chelate-Cram products. We explained this behavior by a chelation between titanium and the β-sulfur present in the substrates.

Owing to the success of aldol-type chain elongations on amino acid derived aldehydes, we were interested in additionally applying aldo-pentoses for the preparation of fluorinated heptoses, which constitute efficient heptosyl-transferase inhibitors[140] in LPS biosynthesis. To this end, we investigated the aldol additions of diisopropylidene-aldehydo-D-arabinose as well as aldehydo-D-arabinose tetraacetate (Scheme 84).

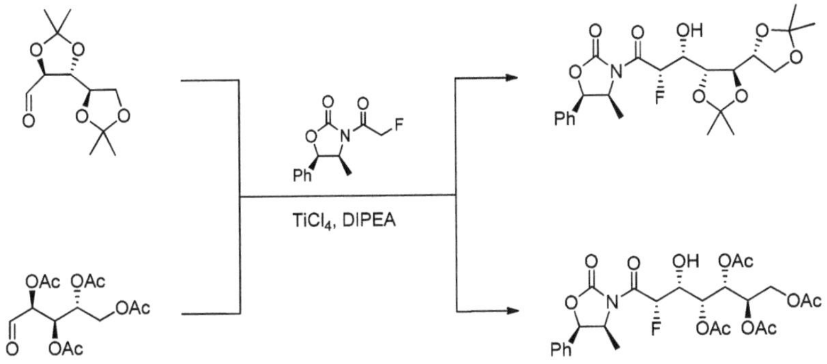

Scheme 84. Aldol additions of D-arabinose derived aldehydes.

Unfortunately, the conversion and yield (~5%) for diisopropylidene protected arabinose were very low. Interestingly, the application of TMEDA in this case resulted only in trace amounts of product. However, the more reactive open chained tetraacetate furnished a yield of ~60% in an initial experiment, albeit low diastereoselectivity was encountered. Thus, we were confident that a corresponding TMEDA mediated reaction would result in good yields and selectivity. Unfortunately, this approach was impeded by the difficult handling of arabinose tetraacetate. Crystallization from acetone/Et$_2$O/hexanes = 2/1/3[141] furnished unreasonably low amounts of pure aldehyde in our hands. Column chromatography resulted in complete decomposition of this labile aldehyde and the utilization of crude samples did not provide any desired aldol products. We reasoned that residual thiol species resulting from the preparation of protected aldehydo-pentoses inhibit the Ti mediated aldolization (Scheme 85).

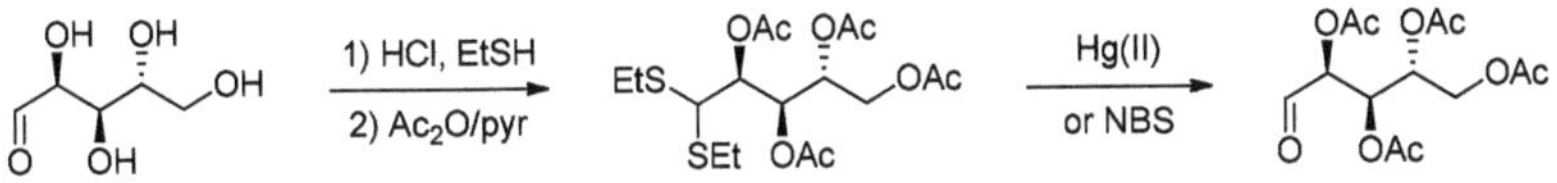

Scheme 85. Preparation of aldehydo-D-arabinose tetraacetate.

Owing to the high lability of the acetyl protected aldehyde precursor, it would be desirable to apply other protecting groups (All, Bn), in order to establish a reliable protocol for the preparation of fluorinated heptoses, which is a matter of future research in our group.

3 Experimental part

3.1 General methods

Oxygen for photochemical oxidations and ozonisation was generated with an Anseros SEP-100 oxygen concentrator (range: 80-93% O_2). NMR spectra were recorded on a Bruker Avance DPX 400 and DPX 600 spectrometer. The chemical shift (δ) is given in parts per million [ppm]. Multiplicities are abbreviated as follows: singlet (s), doublet (d), triplet (t), quadruplet (q), broad signal (brs), multiplet (m), signal shows conformers (cf). Spectra were recorded at 298 K using $CDCl_3$, D_2O or MeOD for referencing. MS experiments were performed in the ESI mode on a Finnigan MAT 900 or Bruker maxis HD spectrometer. IR spectra were recorded on an ELMER FT-IR spectrometer using an ATR unit. For chromatography Merck silica gel 60 (0.004–0.063 mm) was used. For TLC monitoring Merck plates (silica gel 60 F254) were used which were stained by treatment with ninhydrin (0.3 g in 100 ml BuOH, 3 ml AcOH), anis aldehyde (0.5 ml in 50 ml AcOH, 1 ml H_2SO_4), ammonium molybdate (4.8 g + 0.2 g $Ce(SO_4)_2$ in 100 ml H_2SO_4 10%), or $KMnO_4$ (0.5% in H_2O) solutions and charring with a heat gun. UV detection was performed at 254 nm using an UVAC-60 neolab lamp. Acetone, DCM, HE, EA, MeOH and EtOH were distilled before use. Other solvents and chemicals were purchased in reagent grade. Dry DCM (stabilized with 0.2 % EtOH) was prepared by distillation over P_4O_{10} and stored over molecular sieves 4 Å. Other dry solvents were purchased.

Amine bases were freshly distilled from NaH or CaH_2. Optical rotations were measured on a Perkin–Elmer Polarimeter 341 at 589 nm and 20 °C.

3.2 General procedures[1,2]

Method A: Acidic deacetylation, ozonolysis. Synthesis of 6a-c

Acetyl chloride (AcCl) was added to dry MeOH under argon and stirred at room temperature for 15 min. The resulting methanolic HCl solution was added to a solution of the azide in dry MeOH under argon and stirred at room temperature for 16-24 h as judged by TLC (DCM/MeOH = 6/1). To avoid intramolecular 1,4-addition, powdered MS 4Å was then added and the reaction mixture was stirred vigorously for 20 minutes. The molecular sieve was filtered off and the filtrate diluted with dry MeOH and a few ml of dry DCM. The resulting solution was cooled to -78°C and ozone was bubbled through the reaction mixture until a blue color persisted, whereupon air was bubbled through the solution until the blue color vanished. PPh_3 was then added and the reaction mixture was allowed to warm to room temperature overnight. The solution was concentrated under reduced pressure and purification by silica gel chromatography afforded sugar azides **6a-c** as light yellow viscous oils.

Method B: Azide reduction. Synthesis of 7a-7c

The sugar azide was dissolved in pyr/Ac_2O = 1/1 under argon and a catalytic amount of DMAP was added. The resulting reaction mixture was stirred at room temperature for 16 h and then evaporated to dryness. The crude reaction product was redissolved in dry MeCN under argon and DTT followed by DIPA were added. The resulting solution was stirred at room temperature for 2 h and then evaporated to dryness. The residue was redissolved in pyr/Ac_2O = 1/1 under argon and a catalytic amount of DMAP was added. The resulting reaction mixture was stirred at room temperature for 16 h and then evaporated to dryness.

Purification by silica gel chromatography afforded peracetylated amino sugars 7a-c as colorless to light yellow viscous oils.

Method C: Zemplén saponification. Synthesis of 8a-c, 28-31

To a solution of the peracetylated amino sugar in dry MeOH a catalytic amount of NaOMe was added under argon and stirred at room temperature for 2-3 h as judged by TLC (acetone/iPrOH/H_2O = 5/4/1). Then a small amount of acidic ion exchange resin was added and the reaction mixture was stirred for additional 10 min at room temperature. After filtration, the solution was evaporated to dryness and the residue was redissolved in water, washed three times with ethyl acetate EA and evaporated to dryness. The free carbohydrates obtained in most cases needed no further purification.

Method D: Aldol addition. Synthesis of 15-19, 32-33, 42

A stirred solution of (4S,5R)-3-(2-fluoroacetyl)-4-methyl-5-phenyloxazolidin-2-one in dry DCM under an argon atmosphere was cooled to -78 °C and $TiCl_4$ followed by TMEDA were added. The resulting dark brown solution was stirred for 2 h at -78 °C whereupon the aldehyde dissolved in dry DCM was added. The reaction mixture was warmed to -50 °C and was then allowed to slowly warm to -10 °C over 3 h. The reaction was quenched by the addition of saturated ammonium chloride solution and the precipitate formed was filtered over a pad of celite and rinsed with DCM. The phases of the filtrate were separated and the aqueous phase was extracted three times with DCM. The combined organic extracts were dried over anhydrous $MgSO_4$ and filtered. After removal of the solvent under reduced pressure the crude product was purified by flash column chromatography to yield the fluorohydrins as white crystalline solids.

Method E: Oxazolidinone, acetonide cleavage. Synthesis of 20-23

A stirred solution of the fluorohydrin in dry MeOH under an argon atmosphere was cooled to the temperature stated and NaOMe was added. The resulting solution was stirred for 30 min whereupon acidic ion exchange resin was added and the reaction mixture was allowed to warm to room temperature and stirred for the time stated. Afterwards, the reaction mixture was filtered and evaporated to dryness. The crude product was purified by flash column chromatography.

Method F: DIBAL reduction, Boc cleavage. Synthesis of 24-27

A stirred solution of the methyl ester in dry THF under an argon atmosphere was cooled to -78°C and DIBAL (1 M in toluene) was added dropwise. The resulting solution was stirred for 1.5 h at -78°C whereupon the reaction was quenched by the addition of saturated sodium, potassium tartrate solution and allowed to warm to room temperature. The resulting biphasic mixture was diluted with EA and stirred for 3 h. Afterwards the phases were separated and the aqueous phase was extracted three times with EA. The combined organic extracts were dried over anhydrous $MgSO_4$ and filtered. After removal of the solvent under reduced pressure the crude product was dissolved in a mixture of pyr/Ac_2O = 1/1 under an argon atmosphere and a catalytic amount of DMAP was added. The resulting solution was stirred at room temperature over night and then concentrated under reduced pressure and co-evaporated with toluene. The crude reaction product was redissolved in a mixture of TFA/DCM = 1/3 and stirred for 1.5 h at ambient temperature. The reaction mixture was then evaporated to dryness and the crude product redissolved in MeOH. The resulting solution was treated with basic ion exchange resin under stirring until pH = 6 was reached. After filtration, and evaporation of the solvent the crude product was further treated as stated.

Method G: Reductive auxiliary cleavage. Synthesis of 34-35

A stirred solution of the fluoro-hydrin in dry THF under an argon atmosphere was cooled to 0 °C and dry MeOH followed by $LiBH_4$ were added. The reaction mixture was stirred for 45 min at 0 °C and was then quenched by the addition of saturated ammonium chloride solution. The phases were separated and the aqueous phase was extracted three times with DCM. The combined organic extracts were dried over anhydrous $MgSO_4$ and filtered. After removal of the solvent under reduced pressure the crude product was purified by flash column chromatography eluting with HE/EA = 3/2.

Method H: Pummerer rearrangement, acidic deprotection. Synthesis of 40-41

Through a stirred solution of the thiazolidine and TPP in MTBE in a round bottom flask at -78 °C a stream of oxygen was bubbled. A 500 W halogen lamp was illuminated approximately 2 cm above the flask and aluminum foil was used to ensure optimal irradiation which was maintained for 1-2 h whereupon the color of the reaction mixture changed from red to green. Subsequently PPh_3 was added and the reaction mixture was stirred at room temperature for 40 min whereupon the original red coloring was restored. Afterwards, the solvent was evaporated and product and starting material were separated by flash column chromatography eluting with HE/EA = 1/1. The hydroxy-thiazolidine obtained in this way was dissolved in EA and an equal amount of 3 M HCl was added. The resulting biphasic mixture was stirred vigorously for 18 h at room temperature and subsequently evaporated to dryness. The residue was redissolved in 3 M HCl, washed with EA to remove residual TPP and then heated to 100 °C for 3 h. Afterwards, the solvent was removed under reduced pressure and the residue was redissolved in dry MeCN and cooled to 0 °C. TEA was added followed by Ac_2O after 5 min. The resulting solution was stirred at 0 °C for 1 h and then evaporated to dryness. The crude product was purified by flash column chromatography eluting with DCM/MeOH = 19/1.

Procedure for the preparation of the Roush reagent

In a flame dried flask, magnesium turnings (702 mg, 28.87 mmol) were suspended in dry ether (25 ml) under argon and allyl bromide (1.5 ml, 17.32 mmol) was slowly added. After the initial exothermic reaction ceased, the reaction mixture was heated to reflux for 2 h and subsequently allowed to cool to room temperature. In a separate flask, dry ether (25 ml) under argon was cooled to -78 °C. To this flask, the previously prepared Grignard reagent and $B(OMe)_3$ (2.2 ml, 19.25 mmol) in dry ether (25 ml) were slowly added simultaneously. The reaction mixture was stirred for 4 h at -78 °C and subsequently warmed to 0 °C, whereupon HCl 2 M (25 ml) was added. The resulting biphasic mixture was allowed to warm to room temperature stirred vigorously for 1 h. Afterwards, the layers were separated and the aqueous phase was extracted four times with a DCM/ether = 1/5 mixture (25 ml). The combined organic extracts were dried over anhydrous $MgSO_4$ and filtered. After removal of the solvent under reduced pressure the crude product was redissolved in dry ether (50 ml) under argon and treated with (-)-DIPT (3.61 g, 15.4 mmol) at room temperature over night. Subsequently, $MgSO_4$ (1 g) was added and after additional stirring for 20 min the reaction mixture was filtered and evaporated to dryness, furnishing the allyl boronate as a cloudy, white viscous material, which was used without further purification. The purity was estimated to be ~50% according to ^{1}H-NMR analysis.

Procedure for the preparation of flouroacetyl chloride

Ethyl fluoroacetate (13.6 ml, 141 mmol) was dissolved in 200 ml of a mixture of EtOH/H$_2$O = 9/1 and NaOH (6.72 g, 168 mmol) was added. A white precipitate slowly starts to form and after stirring for 20 h at room temperature the solvent was removed under reduced pressure. The sodium fluoroacetate obtained was redissolved in 120 ml of HCl (3 M) and the aqueous solution was saturated with NaCl and then extracted four times with Et$_2$O (100 ml). The combined organic extracts were dried over anhydrous MgSO$_4$, filtered and evaporated to dryness. The obtained fluoroacetic acid (10.4 g, 133 mmol, 95%) is essentially pure and was directly added to PCl$_5$ (30.6 g, 147 mmol) in a flask equipped with a reflux condenser under vigorous stirring and cooling (Caution! Strong exothermic reaction!). After the initial reaction subsided, the reaction mixture was heated at 80 °C for 1 h. Afterwards the fluoroacetyl chloride was directly distilled from the reaction mixture using a short distillation column (b.p. 70-71 °C); yield: 12 g, (80%). The obtained fluoroacetyl chloride contained trace amounts of POCl$_3$ and was used directly without further purification.

Procedure for epimerization of L-threonine to D-*allo*-threonine

L-threonine related Garner's aldehyde (1.5 g, 6.17 mmol) was dissolved in 120 ml of THF and epimerized with LiOH (15 mg, 0.62 mmol). The resulting solution was heated to reflux over night and then evaporated to dryness. The crude product was redissolved in 50 ml of Et$_2$O, washed with water and brine, dried over anhydrous MgSO$_4$, filtered and concentrated under reduced pressure.

The diastereomers were separated by flash column chromatography eluting with HE/EA = 19/1; yield: 495 mg (33%), 990 mg (66%) starting material.

3.4 Experimental procedures and data for key intermediates and final products[1,2]

2-azido-2-deoxy-D-glycero-D-ido-heptose (6a)

Azide **5a** (157 mg, 0.34 mmol) was deacetylated in a methanolic HCl solution using AcCl (73 µl, 1.03 mmol) in 9 ml of dry MeOH according to method A. After ozonolysis, the reaction was quenched with PPh_3 (108 mg, 0.41 mmol). Purification by silica gel chromatography was performed using DCM/MeOH = 6/1 as eluent; yield: 66 mg (mixture of anomers/conformers), (82%). $[\alpha]^D_{20}$ = -26.3° (4.2, H_2O); IR (neat): 3340, 2926, 2117, 1641, 1263, 1042, 813, 737, 631 cm^{-1}; ^1H NMR (D_2O, 600 MHz, 25 °C): (1-H) δ = 4.98 (d, $^3J_{1,2}$ = 8.7 Hz), 5.01 (d, $^3J_{1,2}$ = 4.9 Hz), 5.18 (d, $^3J_{1,2}$ = 1.4 Hz), 5.25 (d, $^3J_{1,2}$ = 3.5 Hz), 5.52 (d, $^3J_{1,2}$ = 4.7 Hz), ^{13}C NMR (D_2O, 150 MHz, 25°C): (1-C) δ = 92.4, 93.1, 93.3, 94.8, 99.0; HRMS (ESI): calcd. for $C_7H_{13}N_3NaO_6$ $[M + Na]^+$ 258.0702, found 258.0701.

2-azido-2-deoxy-D-threo-L-galacto-octose (6b)

Azide **5b** (155 mg, 0.29 mmol) was deacetylated in a methanolic HCl solution using AcCl (62 µl, 0.87 mmol) in 9 ml of dry MeOH according to method A.

Since the deacetylation product was not completely soluble in MeOH about 1.5 mmol of ozone were bubbled through the suspension. After 1 h at -78°C additional 0.75 mmol of ozone were added. After 1 h at -78°C the reaction was quenched with PPh$_3$ (92 mg, 0.35 mmol). Purification by silica gel chromatography was performed using DCM/MeOH = 6/1 as eluent; yield: 55 mg (mixture of anomers, α/β = 1/2), (71%), 3 mg (3%) of not ozonolyzed deacetylation product were recovered. $[α]^D_{20}$ = +17.9° (2.9, H$_2$O); IR (neat): 3341, 2927, 2118, 1591, 1350, 1064, 770, 630 cm^{-1}; ^{1}H NMR (D$_2$O, 600 MHz, 25 °C): (β-anomer) δ = 3.49 (dd, $^3J_{2,3}$ = 10.6 Hz, $^3J_{1,2}$ = 8.1 Hz, 1 H, 2-H), 3.61 (dd, $^3J_{5,6}$ = 9.4 Hz, $^3J_{4,5}$ = 0.9 Hz, 1 H, 5-H), 3.71 (m, 3 H, 8a-H, 8b-H, 3-H), 3.83 (dd, $^3J_{6,7}$ = 1.5 Hz, $^3J_{5,6}$ = 9.4 Hz, 1 H, 6-H), 3.92 (ddd, $^3J_{6,7}$ = 1.5 Hz, $^3J_{7,8a}$ = 5.6 Hz, $^3J_{7,8b}$ = 7.2 Hz, 1 H, 7-H), 4.09 (dd, $^3J_{4,5}$ = 0.9 Hz, $^3J_{3,4}$ = 3.4 Hz, 1 H, 4-H), 4.65 (d, $^3J_{1,2}$ = 8.1 Hz, 1 H, 1-H), (α-anomer) δ = 3.71 (m, 3 H, 8a-H, 8b-H, 2-H), 3.81 (dd, $^3J_{5,6}$ = 9.6 Hz, $^3J_{6,7}$ = 1.7 Hz, 1 H, 6-H), 3.87 (ddd, $^3J_{6,7}$ = 1.7 Hz, $^3J_{7,8b}$ = 5.4 Hz, $^3J_{7,8a}$ = 7.3 Hz, 1 H, 7-H), 4.02 (dd, $^3J_{3,4}$ = 3.2 Hz, $^3J_{2,3}$ = 10.8 Hz, 1 H, 3-H), 4.04 (dd, $^3J_{4,5}$ = 0.9 Hz, $^3J_{5,6}$ = 9.6 Hz, 1 H, 5-H), 4.18 (dd, $^3J_{4,5}$ = 0.9 Hz, $^3J_{3,4}$ = 3.2 Hz, 1 H, 4-H), 5.37 (d, $^3J_{1,2}$ = 3.8 Hz, 1 H, 1-H), ^{13}C NMR (D$_2$O, 150 MHz, 25°C): (β-anomer) δ = 63.0 (8-C), 64.7 (2-C), 67.0 (4-C), 67.4 (6-C), 70.0 (7-C), 72.0 (3-C), 73.0 (5-C), 95.6 (1-C), (α-anomer): 60.5 (2-C), 63.0 (8-C), 67.7 (6-C), 67.9 (4-C), 68.4 (3-C), 68.4 (5-C), 70.1 (7-C), 91.4 (1-C); HRMS (ESI): calcd. for C$_8$H$_{15}$N$_3$NaO$_7$ [M + Na]$^+$ 288.0808, found 288.0791.

2-azido-2-deoxy-D-erythro-L-galacto-octose (6c)

Azide **5c** (102 mg, 0.19 mmol) was deacetylated in a methanolic HCl solution using AcCl (41 μl, 0.58 mmol) in 6 ml of dry MeOH according to method A.

After ozonolysis the reaction was quenched with PPh$_3$ (60 mg, 0.23 mmol). Purification by silica gel chromatography was performed using DCM/MeOH= 6/1 as eluent; yield: 38 mg (mixture of anomers, α/β = 1/2), (75%). [α]$^D_{20}$ = -35.6° (12.9, H$_2$O); IR (neat): 3339, 2923, 2121, 1641, 1252, 1017, 723, 633 cm^{-1}; ^{1}H NMR (D$_2$O, 600 MHz, 25°C): (β-anomer) δ = 3.52 (dd, $^3J_{1,2}$ = 8.1 Hz, $^3J_{2,3}$ = 10.4 Hz, 1 H, 2-H), 3.66 (dd, $^3J_{7,8a}$ = 6.5 Hz, $^2J_{8a,8b}$ = 11.8 Hz, 1 H, 8a-H), 3.67 (dd, $^3J_{3,4}$ = 3.3 Hz, $^3J_{2,3}$ = 10.4 Hz, 1 H, 3-H), 3.69 (dd, $^3J_{4,5}$ = 0.9 Hz, $^3J_{5,6}$ = 5.3 Hz, 1 H, 5-H), 3.77 (dd, $^3J_{7,8b}$ = 3.3 Hz, $^2J_{8a,8b}$ = 11.8 Hz, 1 H, 8b-H), 3.82 (ddd, $^3J_{7,8b}$ = 3.3 Hz, $^3J_{6,7}$ = 5.9 Hz, $^3J_{7,8a}$ = 6.5 Hz, 1 H, 7-H), 3.94 (dd, $^3J_{5,6}$ = 5.3 Hz, $^3J_{6,7}$ = 5.9 Hz, 1 H, 6-H), 4.04 (dd, $^3J_{3,4}$ = 3.3 Hz, $^3J_{4,5}$ = 0.9 Hz, 1 H, 4-H), 4.65 (d, $^3J_{1,2}$ = 8.1 Hz, 1 H, 1-H), (α-anomer) δ = 3.67 (dd, $^3J_{7,8a}$ = 6.5 Hz, $^2J_{8a,8b}$ = 11.8 Hz, 1 H, 8a-H), 3.77 (dd, $^3J_{7,8b}$ = 3.3 Hz, $^2J_{8a,8b}$ = 11.8 Hz, 1 H, 8b-H), 3.73 (dd, $^3J_{1,2}$ = 3.8 Hz, $^3J_{2,3}$ = 10.7 Hz, 1 H, 2-H), 3.80 (ddd, $^3J_{7,8b}$ = 3.3 Hz, $^3J_{6,7}$ = 5.9 Hz, $^3J_{7,8a}$ = 6.5 Hz, 1 H, 7-H), 3.93 (dd, $^3J_{5,6}$ = 5.3 Hz, $^3J_{6,7}$ = 5.9 Hz, 1 H, 6-H), 4.01 (dd, $^3J_{3,4}$ = 3.1 Hz, $^3J_{2,3}$ = 10.7 Hz, 1 H, 3-H), 4.13 (dd, $^3J_{3,4}$ = 3.1 Hz, $^3J_{4,5}$ = 0.9 Hz, 1 H, 4-H), 4.13 (dd, $^3J_{4,5}$ = 0.9 Hz, $^3J_{5,6}$ = 5.3 Hz, 1 H, 5-H), 5.42 (d, $^3J_{1,2}$ = 3.8 Hz, 1 H, 1-H), ^{13}C NMR (CDCl$_3$, 150 MHz, 25°C): (β-anomer) δ = 61.9 (8-C), 64.5 (2-C), 69.5 (4-C), 70.8 (7-C), 71.7 (3-C), 72.3 (6-C), 73.3 (5-C), 95.6 (1-C), (α-anomer) δ = 60.3 (2-C), 62.1 (8-C), 68.1 (3-C), 68.4 (5-C), 70.7 (4-C), 70.8 (7-C), 72.5 (6-C), 91.3 (1-C); HRMS (ESI): calcd. for C$_8$H$_{15}$N$_3$NaO$_7$ [M + Na]$^+$ 288.0808, found 288.0808.

2-acetamido-1,3,4,6,7-penta-O-acetyl-2-deoxy-D-glycero-D-ido-heptose (7a)

Sugar azide **6a** (31 mg, 0.13 mmol) was peracetylated and reduced with DTT (82 mg, 0.53 mmol) and DIPA (1 ml) in 4 ml of dry MeCN according to method B. Purification by silica gel chromatography was performed using HE/EA = 1/3 as eluent; yield: 20 mg, ($^4C_1$-pyranoid form, mixture of anomers, α/β = 3/2), (33%), 16 mg (β-furanoid form/$^1C_4$ α-pyranoid form = 5/2), (26%). ^{1}H NMR (CDCl$_3$, 600 MHz, 25°C): ($^4C_1$-pyranoid form), (α-anomer) δ = 2.01, 2.02, 2.05, 2.10, 2.12, 2.14 (6s, 18 H, 6 Ac), 4.10 (dd, $^3J_{6,7a}$ = 4.6 Hz, $^2J_{7a,7b}$ = 12.4 Hz, 1 H, 7a-H), 4.33 (dddd, $^4J_{2,4}$ = 1.0 Hz, $^3J_{1,2}$ = 1.8 Hz, $^3J_{2,3}$ = 3.1 Hz, $^3J_{2,2\text{-}NH}$ = 9.8 Hz, 1 H, 2-H), 4.38 (ddd, $^4J_{5,1}$ = 0.6 Hz, $^3J_{4,5}$ = 1.9 Hz, $^3J_{5,6}$ = 9.9 Hz, 1 H, 5-H), 4.46 (dd, $^3J_{6,7b}$ = 2.4 Hz, $^2J_{7a,7b}$ = 12.4 Hz, 1 H, 7b-H), 4.84 (ddd, $^4J_{3,1}$ = 1.1 Hz, $^3J_{2,3}$ = 3.1 Hz, $^3J_{3,4}$ = 3.1 Hz, 1 H, 3-H), 5.08 (dddd, $^5J_{4,1}$ = 0.7 Hz, $^4J_{2,4}$ = 1.0 Hz, $^3J_{4,5}$ = 1.9 Hz, $^3J_{3,4}$ = 3.1 Hz, 1 H, 4-H), 5.12 (ddd, $^3J_{6,7b}$ = 2.4 Hz, $^3J_{6,7a}$ = 4.6 Hz, $^3J_{5,6}$ = 9.9 Hz, 1 H, 6-H), 5.92 (dddd, $^4J_{5,1}$ = 0.6 Hz, $^5J_{4,1}$ = 0.7 Hz, $^4J_{3,1}$ = 1.1 Hz, $^3J_{1,2}$ = 1.8 Hz, 1 H, 1-H), 6.05 (d, $^3J_{2,2\text{-}NH}$ = 9.8 Hz, 1 H, NH), ($^4C_1$-pyranoid form), (β-anomer) δ = 2.01, 2.05, 2.09, 2.09, 2.10, 2.17, (6s, 18 H, 6 Ac), 4.13 (dd, $^3J_{6,7a}$ = 5.0 Hz, $^2J_{7a,7b}$ = 12.4 Hz, 1 H, 7a-H), 4.20 (dd, $^3J_{4,5}$ = 1.8 Hz, $^3J_{5,6}$ = 9.8 Hz, 1 H, 5-H), 4.35 (dddd, $^4J_{2,4}$ = 1.0 Hz, $^3J_{1,2}$ = 2.1 Hz, $^3J_{2,3}$ = 2.9 Hz, $^3J_{2,2\text{-}NH}$ = 9.7 Hz, 1 H, 2-H), 4.43 (dd, $^3J_{6,7b}$ = 2.3 Hz, $^2J_{7a,7b}$ = 12.4 Hz, 1 H, 7b-H), 4.97 (ddd, $^4J_{2,4}$ = 1.0 Hz, $^3J_{3,4}$ = 2.9 Hz, $^3J_{2,3}$ = 2.9 Hz, 1 H, 4-H), 5.00 (dd, $^3J_{2,3}$ = 2.9 Hz, $^3J_{3,4}$ = 2.9 Hz, 1 H, 3-H), 5.18 (ddd, $^3J_{6,7b}$ = 2.3 Hz, $^3J_{6,7a}$ = 5.0 Hz, $^3J_{5,6}$ = 9.8 Hz, 1 H, 6-H), 5.94 (d, $^3J_{1,2}$ = 2.1 Hz, 1 H, 1-H), 6.10 (d, $^3J_{2,2\text{-}NH}$ = 9.7 Hz, 1 H, 2-NH), (β-furanoid form) δ = 2.00, 2.04, 2.04, 2.10, 2.12, 2.16 (6s, 18 H, 6 Ac), 4.10 (dd, $^3J_{6,7a}$ = 6.7 Hz, $^2J_{7a,7b}$ = 12.4 Hz, 1 H, 7a-H), 4.37 (dd, $^3J_{6,7b}$ = 3.0 Hz, $^2J_{7a,7b}$ = 12.4 Hz, 1 H, 7b-H), 4.55 (ddd, $^3J_{1,2}$ = 3.1 Hz, $^3J_{2,3}$ = 5.1 Hz, $^3J_{2,2\text{-}NH}$ = 8.1 Hz, 1 H, 2-H), 4.59 (dd, $^3J_{4,5}$ = 5.1 Hz, $^3J_{3,4}$ = 6.8 Hz, 1 H, 4-H), 5.13 (ddd, $^3J_{6,7b}$ = 3.0 Hz, $^3J_{5,6}$ = 5.1 Hz, $^3J_{6,7a}$ = 6.7 Hz, 1 H, 6-H), 5.32 (dd, $^3J_{3,4}$ = 2.9 Hz, $^3J_{2,3}$ = 2.9 Hz, 1 H, 3-H), 5.36 (dd, $^3J_{4,5}$ = 5.1 Hz, $^3J_{5,6}$ = 9.8 Hz, 1 H, 5-H), 5.93 (d, $^3J_{2,2\text{-}NH}$ = 8.1 Hz, 1 H, 2-NH), 6.03 (d, $^3J_{1,2}$ = 2.1 Hz, 1 H, 1-H), ($^1C_4$ α-pyranoid form) δ = 1.95, 2.01, 2.07, 2.12, 2.13, 2.22 (6s, 18 H, 6 Ac), 4.11 (dd, $^3J_{4,5}$ = 1.5 Hz, $^3J_{5,6}$ =

9.6 Hz, 1 H, 5-H), 4.14 (dd, $^3J_{6,7a}$ = 5.1 Hz, $^2J_{7a,7b}$ = 12.3 Hz, 1 H, 7a-H), 4.42 (dd, $^3J_{6,7b}$ = 2.4 Hz, $^2J_{7a,7b}$ = 12.3 Hz, 1 H, 7b-H), 4.54 (ddd, $^3J_{2,3}$ = 3.4 Hz, $^3J_{1,2}$ = 9.3 Hz, $^3J_{2,2\text{-NH}}$ = 9.3 Hz, 1 H, 2-H), 5.04 (dd, $^3J_{4,5}$ = 1.5 Hz, $^3J_{3,4}$ = 3.6 Hz, 1 H, 4-H), 5.15 (ddd, $^3J_{6,7b}$ = 2.4 Hz, $^3J_{6,7a}$ = 5.1 Hz, $^3J_{5,6}$ = 9.6 Hz, 1 H, 6-H), 5.16 (dd, $^3J_{2,3}$ = 3.4 Hz, $^3J_{3,4}$ = 3.6 Hz, 1 H, 3-H), 5.47 (d, $^3J_{2,2\text{-NH}}$ = 9.3 Hz, 1 H, 2-NH), 5.84 (d, $^3J_{1,2}$ = 9.3 Hz, 1 H, 1-H), ^{13}C NMR (CDCl$_3$, 150 MHz, 25°C): ($^4C_1$-pyranoid form), (α-anomer) δ = 20.6, 20.7, 20.7, 20.7, 20.8, 23.2 (6 CH$_3$), 45.7 (2-C), 62.2 (7-C), 65.0 (4-C), 65.2 (5-C), 67.1 (3-C), 67.2 (6-C), 91.8 (1-C), 168.1, 168.5, 168.8, 168.8, 169.7, 170.3 (6 CO-Ac), ($^4C_1$-pyranoid form), (β-anomer) δ = 20.6, 20.7, 20.7, 20.6, 20.8, 23.3 (6 Ac), 47.0 (2-C), 62.4 (7-C), 64.2 (4-C), 67.0 (6-C), 68.7 (3-C), 72.3 (5-C), 90.7 (1-C), 168.2, 168.3, 168.6, 169.4, 169.8, 170.6 (6 CO-Ac), (β-furanoid form) δ = 20.6, 20.7, 20.8, 20.9, 21.1, 23.1 (6 Ac), 60.0 (2-C), 61.8 (7-C), 69.2 (5-C), 70.1 (6-C), 74.8 (4-C), 78.3 (3-C), 98.5 (1-C), 169.5, 169.8, 170.0, 170.1, 170.3, 170.7 (6 CO-Ac), ($^1C_4$ α-pyranoid form) δ = 20.7, 20.7, 20.7, 20.9, 21.0, 23.2 (6 Ac), 47.3 (2-C), 62.5 (7-C), 64.8 (4-C), 67.3 (6-C), 70.0 (3-C), 71.1 (5-C), 91.4 (1-C), 168.8, 169.4, 169.6, 169.6, 170.0, 170.6 (6 CO-Ac); HRMS (ESI): calcd. for C$_{19}$H$_{27}$NNaO$_{12}$ [M + Na]$^+$ 484.1431, found 484.1419 and 484.1425.

2-acetamido-1,3,4,6,7,8-hexa-O-acetyl-2-deoxy-D-threo-L-galacto-octose (7b)

Sugar azide **6b** (55 mg, 0.21 mmol) was peracetylated and reduced with DTT (128 mg, 0.83 mmol) and DIPA (1.5 ml) in 6 ml of dry MeCN according to method B. Purification by silica gel chromatography was performed using HE/EA = 1/4 as eluent; yield 73 mg (mixture of anomers, α/β = 1/1), (66%). ^{1}H NMR (CDCl$_3$, 400 MHz, 25°C): (β-anomer) δ = 1.93, 2.00, 2.01, 2.01, 2.04,

2.12, 2.13 (7s, 21 H, 7 Ac), 3.87 (dd, $^3J_{4,5}$ = 1.1 Hz, $^3J_{5,6}$ = 9.5 Hz, 1 H, 5-H), 3.96 (dd, $^2J_{8a,8b}$ = 11.8 Hz, $^3J_{7,8a}$ = 5.8 Hz, 1 H, 8a-H), 4.17 (dd, $^3J_{7,8b}$ = 4.8 Hz, $^2J_{8a,8b}$ = 11.8 Hz, 1 H, 8b-H), 4.36 (ddd, $^3J_{1,2}$ = 9.0 Hz, $^3J_{2,2\text{-NH}}$ = 9.1 Hz, $^3J_{2,3}$ = 11.3 Hz, 1 H, 2-H), 5.12 (dd, $^3J_{3,4}$ = 3.5 Hz, $^3J_{2,3}$ = 11.3 Hz, 1 H, 3-H), 5.30 (d, $^3J_{2,2\text{-NH}}$ = 9.1 Hz, 1 H, 2-NH), 5.35 (m, 3 H, 4-H, 6-H, 7-H), 5.62 (d, $^3J_{1,2}$ = 9.0 Hz, 1 H, 1-H), (α-anomer) δ = 1.94, 2.01, 2.02, 2.10, 2.12, 2.12, 2.15 (7s, 21 H, 7 Ac), 3.94 (dd, $^3J_{7,8a}$ = 7.1 Hz, $^2J_{8a,8b}$ = 11.8 Hz, 1 H, 8a-H), 4.06 (dd, $^3J_{4,5}$ = 0.9 Hz, $^3J_{5,6}$ = 9.7 Hz, 1 H, 5-H), 4.21 (dd, $^3J_{7,8b}$ = 5.6 Hz, $^2J_{8a,8b}$ = 11.8 Hz, 1 H, 8b-H), 4.72 (ddd, $^3J_{1,2}$ = 3.7 Hz, $^3J_{2,2\text{-NH}}$ = 9.3 Hz, $^3J_{2,3}$ = 11.5 Hz, 1 H, 2-H), 5.15 (dd, $^3J_{6,7}$ = 2.0 Hz, $^3J_{5,6}$ = 9.7 Hz, 1 H, 6-H), 5.18 (dd, $^3J_{3,4}$ = 3.3 Hz, $^3J_{2,3}$ = 11.5 Hz, 1 H, 3-H), 5.31 (d, $^3J_{2,2\text{-NH}}$ = 9.3 Hz, 1 H, 2-NH), 5.35 (m, 2 H, 4-H, 7-H), 6.23 (d, $^3J_{1,2}$ = 3.7 Hz, 1 H, 1-H); ^{13}C-NMR (CDCl$_3$, 100 MHz, 25 °C): (β-anomer) δ = 20.5, 20.6, 20.6, 20.7, 20.7, 20.8, 23.7 (7 Ac), 50.1 (2-C), 62.9 (8-C), 65.1 (6-C), 65.6 (4-C), 68.7 (7-C), 70.3 (3-C), 71.6 (5-C), 93.2 (1-C), 169.2, 169.5, 169.9, 170.2, 170.4, 170.7, 171.2 (7 CO-Ac), (α-anomer) δ = 20.4, 20.6, 20.6, 20.6, 20.7, 20.8, 23.6 (7 Ac), 47.1 (2-C), 62.1 (8-C), 67.1 (4-C), 67.4 (6-C), 68.1 (3-C), 68.2 (5-C), 68.2 (7-C), 91.2 (1-C), 168.5, 169.4, 169.5, 169.6, 170.2, 170.4, 170.6 (7 CO-Ac); HRMS (ESI): calcd. for C$_{22}$H$_{31}$NNaO$_{14}$ [M + Na]$^+$ 556.1642, found 556.1633.

2-acetamido-1,3,4,6,7,8-hexa-O-acetyl-2-deoxy-D-erythro-L-galacto-octose (7c)

Sugar azide **6c** (38 mg, 0.14 mmol) was peracetylated and reduced with DTT (88 mg, 0.57 mmol) and DIPA (1 ml) in 4 ml of dry MeCN according to method B. Purification by silica gel chromatography was performed using HE/EA = 1/4 as eluent; yield: 51 mg, (mixture of anomers, α/β = 1/2), (67%). [α]$^D_{20}$ = -12.0°

(1.0, CH$_2$Cl$_2$); ^{1}H NMR (CDCl$_3$, 400 MHz, 25°C):, (β-anomer) δ = 1.94, 2.01, 2.01, 2.07, 2.11, 2.11, 2.23 (7s, 21 H, 7 Ac), 3.90 (dd, $^3J_{4,5}$ = 0.5 Hz, $^3J_{5,6}$ = 7.6 Hz, 1 H, 5-H), 4.17 (dd, $^3J_{7,8a}$ = 5.9 Hz, $^2J_{8a,8b}$ = 11.9 Hz, 1 H, 8a-H), 4.28 (dd, $^3J_{7,8b}$ = 5.1 Hz, $^2J_{8a,8b}$ = 11.9 Hz, 1 H, 8b-H), 4.32 (ddd, $^3J_{1,2}$ = 8.8 Hz, $^3J_{2,2\text{-NH}}$ = 9.4 Hz, $^3J_{2,3}$ = 11.1 Hz, 1 H, 2-H), 4.95 (ddd, $^3J_{6,7}$ = 3.5 Hz, $^3J_{7,8b}$ = 5.1 Hz, $^3J_{7,8a}$ = 5.9 Hz, 1 H, 7-H), 5.14 (dd, $^3J_{3,4}$ = 3.3 Hz, $^3J_{2,3}$ = 11.1 Hz, 1 H, 3-H), 5.32 (d, $^3J_{2,2\text{-NH}}$ = 9.4 Hz, 1 H, 2-NH), 5.46 (dd, $^3J_{6,7}$ = 3.5 Hz, $^3J_{5,6}$ = 7.6 Hz, 1 H, 6-H), 5.52 (dd, $^3J_{4,5}$ = 0.5 Hz, $^3J_{3,4}$ = 3.3 Hz, 1 H, 4-H), 5.65 (d, $^3J_{1,2}$ = 8.8 Hz, 1 H, 1-H), (α-anomer) δ = 1.95, 2.02, 2.02, 2.07, 2.08, 2.18, 2.21 (7s, 21 H, 7 Ac), 4.05 (dd, $^3J_{4,5}$ = 1.0 Hz, $^3J_{5,6}$ = 6.6 Hz, 1 H, 5-H), 4.14 (dd, $^3J_{7,8a}$ = 6.3 Hz, $^2J_{8a,8b}$ = 12.0 Hz, 1 H, 8a-H), 4.28 (dd, $^3J_{7,8b}$ = 5.1 Hz, $^2J_{8a,8b}$ = 12.0 Hz, 1 H, 8b-H), 4.72 (ddd, $^3J_{1,2}$ = 3.7 Hz, $^3J_{2,2\text{-NH}}$ = 9.2 Hz, $^3J_{2,3}$ = 11.6 Hz, 1 H, 2-H), 5.00 (ddd, $^3J_{6,7}$ = 4.1 Hz, $^3J_{7,8b}$ = 5.1 Hz, $^3J_{7,8a}$ = 6.3 Hz, 1 H, 7-H), 5.16 (dd, $^3J_{3,4}$ = 3.3 Hz, $^3J_{2,3}$ = 11.6 Hz, 1 H, 3-H), 5.32 (d, $^3J_{2,2\text{-NH}}$ = 9.2 Hz, 1 H, 2-NH), 5.37 (dd, $^3J_{6,7}$ = 4.1 Hz, $^3J_{5,6}$ = 6.6 Hz, 1 H, 6-H), 5.53 (dd, $^3J_{4,5}$ = 1.0 Hz, $^3J_{3,4}$ = 3.3 Hz, 1 H, 4-H), 6.19 (d, $^3J_{1,2}$ = 3.7 Hz, 1 H, 1-H); ^{13}C-NMR (CDCl$_3$, 100 MHz, 25 °C): (β-anomer) δ = 20.6, 20.6, 20.7, 20.7, 20.8, 20.9, 23.4 (7 Ac), 50.2 (2-C), 61.1 (8-C), 66.2 (4-C), 69.2 (7-C), 69.9 (6-C), 70.3 (3-C), 73.3 (5-C), 93.2 (1-C), 169.6, 169.8, 170.1, 170.2, 170.5, 170.6, 170.9 (7 CO-OAc), (α-anomer) δ = 20.6, 20.6, 20.7, 20.7, 20.8, 20.9, 23.2 (7 Ac), 46.9 (2-C), 61.4 (8-C), 67.0 (4-C), 68.1 (3-C), 69.0 (5-C), 69.5 (6-C), 69.6 (7-C), 91.6 (1-C), 169.7, 169.8, 170.9, 170.3, 170.5, 170.52, 170.5 (7 CO-OAc); HRMS (ESI): calcd. for C$_{22}$H$_{31}$NNaO$_{14}$ [M + Na]$^+$ 556.1642, found 556.1638.

2-acetamido-2-deoxy-D-glycero-D-ido-heptose (8a)

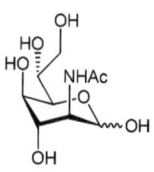

Peracetylated amino sugar **7a** (20 mg, 0.04 mmol) was deacetylated according to method **C** in 3 ml of dry MeOH; yield: 11 mg, (100%). $[\alpha]^D_{20}$ = -12.8° (2.5, H$_2$O); ^{1}H NMR (D$_2$O, 600 MHz, 25 °C): (1-H) δ = 4.92 (d, $^3J_{1,2}$ = 8.9 Hz), 5.05 (d, $^3J_{1,2}$ = 3.0 Hz), 5.15 (d, $^3J_{1,2}$ = 3.8 Hz), 5.19 (d, $^3J_{1,2}$ = 1.9 Hz), 5.45 (d, $^3J_{1,2}$ = 4.9 Hz), ^{13}C NMR (D$_2$O, 150 MHz, 25°C): (1-C) δ = 91.4, 91.8, 92.9, 93.5, 93.6; HRMS (ESI): calcd. for C$_9$H$_{17}$NNaO$_7$ [M+Na]$^+$ 274.0903, found 274.0905.

2-acetamido-2-deoxy-D-threo-L-galacto-octose (8b)

Peracetylated amino sugar **7b** (44 mg, 0.08 mmol) was deacetylated according to method **C** in 4 ml of dry methanol; yield: 23 mg, (mixture of anomers, α/β = 1/1), (100%). $[\alpha]^D_{20}$ = -30° (5.0, H$_2$O), ^{1}H NMR (MeOD, 600 MHz, 25°C): (β-anomer) δ = 2.00 (s, 3 H, NHAc), 3.52 (dd, $^3J_{4,5}$ = 1.0 Hz, $^3J_{5,6}$ = 9.0 Hz, 1 H, 5-H), 3.58 (dd, $^3J_{3,4}$ = 3.4 Hz, $^3J_{2,3}$ = 10.8 Hz, 1 H, 3-H), 3.65 (m, 2 H, 8a-H, 8b-H), 3.88 (m, 3 H, 2-H, 6-H, 7-H), 4.04 (dd, $^3J_{4,5}$ = 1.0 Hz, $^3J_{3,4}$ = 3.4 Hz, 1 H, 4-H), 4.55 (d, $^3J_{1,2}$ = 8.4 Hz, 1 H, 1-H), (α-anomer) δ = 2.00 (s, 3 H, NHAc) 3.65 (m, 2 H, 8a-H, 8b-H), 3.83 (dd, $^3J_{3,4}$ = 3.2 Hz, $^3J_{2,3}$ = 10.9 Hz, 1 H, 3-H), 3.83 (ddd, $^3J_{6,7}$ = 1.6 Hz, $^3J_{7,8a}$ = 6.5 Hz, $^3J_{7,8b}$ = 6.5 Hz, 1 H, 7-H), 3.85 (dd, $^3J_{6,7}$ = 1.6 Hz, $^3J_{5,6}$ = 9.2 Hz, 1 H, 6-H), 4.03 (dd, $^3J_{4,5}$ = 1.2 Hz, $^3J_{5,6}$ = 9.2 Hz, 1 H, 5-H), 4.10 (dd, $^3J_{4,5}$ = 1.2 Hz, $^3J_{3,4}$ = 3.2 Hz, 1 H, 4-H), 4.22 (dd, $^3J_{1,2}$ = 3.7 Hz, $^3J_{2,3}$ = 10.9 Hz, 1 H, 2-H), 5.14 (d, $^3J_{1,2}$ = 3.7 Hz, 1 H, 1-H), ^{13}C NMR (MeOD, 150 MHz, 25°C): (β-anomer) δ = 22.9 (NHAc), 55.9 (2-C), 64.8 (8-C), 68.7 (4-C), 69.1 (6-C), 71.7 (7-C), 73.6 (3-C), 74.8 (5-C), 97.6 (1-C), 174.7 (CO-NHAc), (α-anomer) δ = 22.7 (NHAc), 52.1 (2-C), 64.8 (8-C), 69.5 (6-C), 69.6 (4-C), 69.9 (3-C), 70.0 (5-C), 71.8 (7-C), 93.0 (1-C), 174.0 (CO-NHAc); HRMS (ESI): calcd. for C$_{10}$H$_{19}$NNaO$_8$ [M + Na]$^+$ 304.1008, found 304.1003.

2-acetamido-2-deoxy-D-erythro-L-galacto-octose (8c)

Peracetylated amino sugar **7c** (51 mg, 0.10 mmol) was deacetylated according to method C in 5 ml of dry methanol; yield: 27 mg, (mixture of anomers, α/β = 1/1), (100%). $[\alpha]^D_{20}$ = -31.6° (7.9, H$_2$O); ^{1}H NMR (D$_2$O, 600 MHz, 25°C): (β-anomer) δ = 2.06 (s, 3 H, NHAc), 3.67 (dd, $^3J_{7,8a}$ = 6.7 Hz, $^2J_{8a,8b}$ = 12.0 Hz, 1 H, 8a-H), 3.70 (dd, $^3J_{4,5}$ = 1.1 Hz, $^3J_{5,6}$ = 5.3 Hz, 1 H, 5-H), 3.72 (dd, $^3J_{3,4}$ = 3.3 Hz, $^3J_{2,3}$ = 10.8 Hz, 1 H, 3-H), 3.78 (dd, $^3J_{7,8b}$ = 3.5 Hz, $^2J_{8a,8b}$ = 12.0 Hz, 1 H, 8b-H), 3.83 (ddd, $^3J_{7,8b}$ = 3.5 Hz, $^3J_{6,7}$ = 6.1 Hz, $^3J_{7,8a}$ = 6.7 Hz, 1 H, 7-H), 3.92 (dd, $^3J_{1,2}$ = 8.5 Hz, $^3J_{2,3}$ = 10.8 Hz, 1 H, 2-H), 3.96 (dd, $^3J_{5,6}$ = 5.3 Hz, $^3J_{6,7}$ = 6.1 Hz, 1 H, 6-H), 4.06 (dd, $^3J_{4,5}$ = 1.1 Hz, $^3J_{3,4}$ = 3.3 Hz, 1 H, 4-H), 4.66 (d, $^3J_{1,2}$ = 8.5 Hz, 1 H, 1-H), (α-anomer) δ = 2.06 (s, 3 H, NHAc) 3.67 (dd, $^3J_{7,8a}$ = 6.5 Hz, $^2J_{8a,8b}$ = 11.9 Hz, 1 H, 8a-H), 3.78 (dd, $^3J_{7,8b}$ = 3.5 Hz, $^2J_{8a,8b}$ = 11.9 Hz, 1 H, 8b-H), 3.81 (ddd, $^3J_{7,8b}$ = 3.5 Hz, $^3J_{6,7}$ = 6.2 Hz, $^3J_{7,8a}$ = 6.5 Hz, 1 H, 7-H), 3.92 (dd, $^3J_{3,4}$ = 3.2 Hz, $^3J_{2,3}$ = 10.9 Hz, 1 H, 3-H), 3.94 (dd, $^3J_{5,6}$ = 5.0 Hz, $^3J_{6,7}$ = 6.2 Hz, 1 H, 6-H), 4.12 (dd, $^3J_{4,5}$ = 1.3 Hz, $^3J_{3,4}$ = 3.2 Hz, 1 H, 4-H), 4.14 (dd, $^3J_{4,5}$ = 1.3 Hz, $^3J_{5,6}$ = 5.0 Hz, 1 H, 5-H), 4.18 (dd, $^3J_{1,2}$ = 3.8 Hz, $^3J_{2,3}$ = 10.9 Hz, 1 H, 2-H), 5.28 (d, $^3J_{1,2}$ = 3.8 Hz, 1 H, 1-H), ^{13}C NMR (D$_2$O, 150 MHz, 25°C): (β-anomer) δ = 22.2 (NHAc), 53.5 (2-C), 61.8 (8-C), 69.4 (4-C), 70.1 (7-C), 71.2 (3-C), 72.4 (6-C), 73.2 (5-C), 95.5 (1-C), 175.0 (CO-NHAc), (α-anomer) δ = 22.0 (NHAc), 50.1 (2-C), 62.1 (8-C), 67.5 (3-C), 68.4 (5-C), 70.3 (4-C), 71.0 (7-C), 72.6 (6-C), 91.0 (1-C), 174.7 (CO-NHAc), HRMS (ESI): calcd. for C$_{10}$H$_{19}$NNaO$_8$ [M + Na]$^+$ 304.1008, found 304.1002.

(R,E)-methyl 4-(dibenzylamino)-5-((4-methoxybenzyl)oxy)pent-2-enoate (9a)

PMBO~~~NBn$_2$~~~CO$_2$Me

Oxalyl chloride (164 µl, 1.91 mmol) in dry DCM (20 ml) under argon was cooled to -78 °C and dry DMSO (163 µl, 2.29 mmol) was slowly added. The resulting solution was stirred at -78 °C for 5 min and then treated with (R)-2-(dibenzylamino)-3-((4-methoxybenzyl)oxy)propan-1-ol (500 mg, 1.28 mmol). After stirring at -78 °C for 1 h TEA (517 µl, 3.73 mmol) was added and the reaction mixture was allowed to warm to room temperature and subsequently quenched with HCl 1% (20 ml). The layers were separated and the organic phase was washed with saturated NaHCO$_3$ solution (20 ml). The combined aqueous phases were extracted two times with DCM (20 ml) and the combined organic extracts were dried over anhydrous MgSO$_4$ and filtered. After removal of the solvent under reduced pressure the crude product was redissolved in dry DCM (20 ml) and treated with methyl 2-(triphenylphosphoranylidene)acetate (853 mg, 2.55 mmol). The resulting solution was stirred for 1 h at room temperature and subsequently evaporated to dryness. Purification by flash column chromatography eluting with HE/EA = 4/1 afforded compound **9a** as a yellow oil; yield: 500 mg (88%). ^{1}H NMR (CDCl$_3$, 400 MHz, 25°C): δ = 3.56 (m, 1 H, 4-H), 3.60 (d, ^2J = 13.9 Hz, 2 H, N-CH$_2$-Ph), 3.62 (dd, $^3J_{4,5a}$ = 6.8 Hz, $^2J_{5a,5b}$ = 9.4 Hz, 1 H, 5a-H), 3.74 (dd, $^3J_{4,5b}$ = 5.6 Hz, $^2J_{5a,5b}$ = 9.4 Hz, 1 H, 5b-H), 3.77 (s, 3 H, -COOCH$_3$), 3.77 (d, ^2J = 13.6 Hz, 2 H, N-CH$_2$-Ph), 3.81 (s, 3 H, Ph-OCH$_3$), 4.41 (s, 2 H, O-CH$_2$-Ar), 6.03 (dd, $^3J_{2,3}$ = 15.8 Hz, $^4J_{2,4}$ = 1.3 Hz, 1 H, 2-H), 6.85-6.89 (m, 2 H, o-CH-Ar-PMB), 7.04 (dd, $^3J_{3,4}$ = 6.6 Hz, $^3J_{2,3}$ = 15.8 Hz, 1 H, 3-H), 7.19-7.38 (m, 12 H, CH-Ar), ^{13}C NMR (CDCl$_3$, 100 MHz, 25°C): δ = 51.7 (-COOCH$_3$), 54.7 (N-CH$_2$-Ph), 55.4 (Ar-OCH$_3$), 58.4 (4-C), 70.0 (5-C), 73.0 (O-CH$_2$-Ar), 114.0 (o-CH-Ar-PMB), 123.6 (2-C), 127.1, 128.5, 128.6, 129.4 (CH-Ar), 130.3 (Cq-Ar-OCH$_3$), 139.8 (Cq-Ar-NBn$_2$), 146.4 (3-C), 159.4

(Cq-Ar-OCH$_2$), 166.9 (1-C); HRMS (ESI): calcd. for C$_{28}$H$_{31}$NNaO$_4$ [M + Na]$^+$ 468.2151, found 468.2156.

(R,E)-methyl 5-((tert-butyldiphenylsilyl)oxy)-4-(dibenzylamino)pent-2-enoate (9b)

TBDPSO~~~NBn$_2$~~~CO$_2$Me

Compound **9b** was prepared analogously to **9a** from (S)-methyl 3-((tert-butyldiphenylsilyl)oxy)-2-(dibenzylamino)propanoate (1.40 g, 2.61 mmol). Flash column chromatography was performed eluting with HE/EA = 19/1; yield: 986 mg (67%). For spectroscopic data see Ref. 126.

(R,E)-methyl 4-(dibenzylamino)-5-(pivaloyloxy)pent-2-enoate (9c)

PivO~~~NBn$_2$~~~CO$_2$Me

A solution of compound **9b** (905 mg, 1.61 mmol) in dry THF (8 ml) under argon was treated with TBAF (1 M in THF, 2.1 ml, 2.1 mmol) and the resulting solution was stirred at room temperature for 1 h. The reaction was quenched by adding saturated NaCl solution (10 ml) and the layers were separated. The organic phase was dried over anhydrous MgSO$_4$ and filtered. After removal of the solvent under reduced pressure the crude product was redissolved in a mixture of dry pyridine/DCM = 1/1 (20 ml) and a catalytic amount of DMAP was added. Subsequently PivCl (810 µl, 6.58 mmol) was added dropwise and the resulting solution was stirred for 1 h at room temperature. The reaction was quenched by the addition of water (20 ml) and after separation of the layers the aqueous phase was extracted three times with DCM (20 ml). The combined organic extracts were dried over anhydrous MgSO$_4$ and filtered. After removal of the solvent under reduced pressure the crude product was purified by flash column chromatography eluting with HE/EA = 2/1; yield: 540 mg (82 %). ^{1}H NMR (CDCl$_3$, 400 MHz, 25°C): δ = 1.19 (s, 9 H, 3 CH$_3$), 3.60 (d, ^2J = 13.8 Hz, 2 H, N-CH$_2$-Ph), 3.62 (m, 1 H, 4-H), 3.78 (s, 3 H, -OCH$_3$), 3.82 (d, ^2J = 13.9 Hz, 2 H, N-CH$_2$-Ph), 4.23 (dd, $^3J_{4,5a}$ = 6.3 Hz, $^2J_{5a,5b}$ = 11.4 Hz, 1 H, 5a-H), 4.34 (dd, $^3J_{4,5b}$ = 6.6 Hz, $^2J_{5a,5b}$ = 11.4 Hz, 1 H, 5b-H), 6.00 (dd, $^3J_{2,3}$ = 15.9 Hz, $^4J_{2,4}$ = 1.3 Hz, 1 H, 2-H), 6.98 (dd, $^3J_{3,4}$ = 7.3 Hz, $^3J_{2,3}$ = 15.9 Hz, 1 H, 3-H), 7.21-7.39 (m, 10 H, CH-Ar), ^{13}C NMR (CDCl$_3$, 100 MHz, 25°C): δ = 27.3 (3 CCH$_3$), 38.9 (Cq-Piv), 51.8 (-OCH$_3$), 54.6 (N-CH$_2$-Ph), 57.9 (4-C), 63.3 (5-C), 124.4 (2-C), 127.3, 128.5, 128.6 (CH-Ar), 139.3 (Cq-Ar-NBn$_2$), 144.3 (3-C), 166.5 (1-C), 178.4 (C=O-Piv); HRMS (ESI): calcd. for C$_{25}$H$_{32}$NO$_4$ [M + H]$^+$ 410.2331, found 410.2307.

((2S,3S)-3-((S)-1-(dibenzylamino)-2-((4-methoxybenzyl)oxy)ethyl)oxiran-2-yl)methanol (10)

Compound **9a** (495 mg, 1.11 mmol) was treated according to method F with DIBAL (3.9 ml, 3.9 mmol) furnishing essentially pure allylic alcohol [yield: 458 mg (99%)] which was subsequently epoxidized under the conditions developed by Sharpless *et al.* (-)-Diisopropyl D-tartrate (128 mg, 0.55 mmol) and Ti(OiPr)$_4$ (114 μl, 0.38 mmol) were added to a suspension of powdered molecular sieves 4 Å (300 mg) in dry DCM (20 ml) under argon and cooled to -35 °C. After stirring for 30 min, the allylic alcohol (458 mg, 1.10 mmol) and tBuOOH (5.5 M in nonane, 300 μl, 1.65 mmol) were added and the reaction mixture was stirred for 24 h at -30 °C. The reaction was quenched by the addition of tartaric acid (120 mg/ml, 10 ml) and FeSO$_4$.7H$_2$O (400 mg). After stirring for 90 min at room temperature, the reaction mixture was diluted with DCM (20 ml), dried over

MgSO$_4$ and filtered. After removal of the solvent under reduced pressure, the crude product was purified by flash column chromatography, eluting with HE/EA = 4/1; yield: 252 mg (53%), 188 mg (41%) of starting material recovered. ^{1}H NMR (CDCl$_3$, 400 MHz, 25°C): δ = 1.60 (brs, 1 H, 5-OH), 2.86 (ddd, $^3J_{4,5b}$ = 5.6 Hz, $^3J_{3,4}$ = 6.8 Hz, $^3J_{4,5a}$ = 6.9 Hz, 1 H, 4-H), 2.99 (ddd, $^3J_{1a,2}$ = 2.5 Hz, $^3J_{1b,2}$ = 4.1 Hz, $^3J_{2,3}$ = 2.3 Hz, 1 H, 2-H), 3.22 (dd, $^3J_{2,3}$ = 2.3 Hz, $^3J_{3,4}$ = 6.8 Hz, 1 H, 3-H), 3.58 (dd, $^3J_{4,5a}$ = 6.9 Hz, $^2J_{5a,5b}$ = 9.7 Hz, 1 H, 5a-H), 3.59 (m, 1 H, 1a-H), 3.66 (dd, $^3J_{4,5b}$ = 5.6 Hz, $^2J_{5a,5b}$ = 9.7 Hz, 1 H, 5b-H), 3.81 (s, 3 H, Ph-OCH$_3$), 3.83 (m, 1 H, 1b-H), 3.85, 3.86 (2s, 4 H, 2 N-CH$_2$-Ph), 4.38 (s, 2 H, O-CH$_2$-Ar), 6.85-6.89 (m, 2 H, o-CH-Ar-PMB), 7.18-7.39 (m, 12 H, CH-Ar), ^{13}C NMR (CDCl$_3$, 100 MHz, 25°C): δ = 55.4 (3-C, Ar-OCH$_3$), 55.6 (N-CH$_2$-Ph), 56.0 (2-C), 58.6 (4-C), 61.8 (1-C), 69.4 (5-C), 73.1 (O-CH$_2$-Ar), 114.0 (o-CH-Ar-PMB), 127.0, 128.3, 128.8, 129.3 (CH-Ar), 130.3 (Cq-Ar-OCH$_3$), 140.2 (Cq-Ar-NBn$_2$), 159.4 (Cq-Ar-OCH$_2$); HRMS (ESI): calcd. for C$_{27}$H$_{32}$NO$_4$ [M + H]$^+$ 434.2331, found 434.2309.

(2R,3S)-methyl 3-((S)-2-((tert-butyldiphenylsilyl)oxy)-1-(dibenzylamino)ethyl)oxirane-2-carboxylate (11)

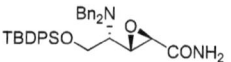

A flame dried, three-necked flask was charged with KOtBu (134 mg, 1.19 mmol) under argon and cooled to -78 °C. Dry ammonia was purged through the flask until condensation of ~20 ml of NH$_3$ was achieved. Compound **9b** (560 mg, 0.99 mmol) in dry THF (2 ml) was added followed by tBuOOH (5.5 M in nonane, 217 μl, 1.19 mmol). The resulting brightly red colored solution was allowed to warm to -40 °C and stirred for 12 h. The ammonia was subsequently left to evaporate at 0 °C and the residue was redissolved in pH 7 phosphate buffer (20 ml) and extracted three times with ether (20 ml). The combined organic extracts were washed with brine, dried over anhydrous MgSO$_4$ and filtered. After removal of the solvent under reduced pressure the crude product was purified by flash column chromatography, eluting with HE/EA = 4/1; yield: 68 mg (8%). ^{1}H NMR (CDCl$_3$, 400 MHz, 25°C): δ = 1.04 (s, 9 H, 3 CCH$_3$), 2.81 (ddd, $^3J_{4,5b}$ = 5.8 Hz, $^3J_{3,4}$ = 5.8 Hz, $^3J_{4,5a}$ = 6.4 Hz, 1 H, 4-H), 3.28 (dd, $^3J_{2,3}$ = 2.1 Hz, $^3J_{3,4}$ = 5.8 Hz, 1 H, 3-H), 3.33 (d, $^3J_{2,3}$ = 2.1 Hz, 1 H, 2-H), 3.77 (d, ^2J = 13.8 Hz, 2 H, N-CH$_2$-Ph), 3.82 (dd, $^3J_{4,5a}$ = 6.4 Hz, $^2J_{5a,5b}$ = 10.6 Hz, 1 H, 5a-H), 3.86 (d, ^2J = 13.6 Hz, 2 H, N-CH$_2$-Ph), 3.89 (dd, $^3J_{4,5b}$ = 5.8 Hz, $^2J_{5a,5b}$ = 9.7 Hz, 1 H, 5b-H), 3.81 (s, 3 H, Ph-OCH$_3$), 3.83 (m, 1 H, 1b-H), 5.35, 5.95 (2 brs, 2 H, -NH$_2$), 7.16-7.43, 7.58-7.66 (m, 20 H, CH-Ar), ^{13}C NMR (CDCl$_3$, 100 MHz, 25°C): δ = 19.3 (Cq-*t*-Bu), 27.0 (3 CCH$_3$) 52.7 (2-C), 55.5 (N-CH$_2$-Ph), 59.0 (3-C), 59.7 (4-C), 62.8 (5-C), 127.1, 128.0, 128.4, 128.8, 130.0, 135.8 (CH-Ar), 133.0, 133.2, 139.8 (Cq-Ar), 171.1 (1-C); HRMS (ESI): calcd. for C$_{35}$H$_{40}$N$_2$NaO$_3$Si [M + Na]$^+$ 587.2706, found 587.2695.

(S)-tert-butyl 4-((R)-1-(benzyloxy)-3-oxopropyl)-2,2-dimethyloxazolidine-3-carboxylate (12)

L-Garner's aldehyde (917 mg, 4.00 mmol) and freshly prepared Roush reagent (2.5 ml, see section 3.2) were added separately to two suspensions of powdered molecular sieves 4 Å (300 mg) in dry ether (15 ml) under argon and stirred vigorously for 30 min at room temperature. The suspension of the aldehyde was cooled to -78 °C and the pre-dried allyl-borate was added drop-wise. The reaction mixture was stirred for 18 h at -78 °C and quenched by the addition of NaOH 1 M (20 ml). After stirring for 1 h at room temperature, the reaction mixture was filtered and the layers were separated. The aqueous phase was extracted three times with EA (30 ml) and the combined organic extracts were washed with brine, dried over anhydrous MgSO₄ and filtered. After removal of the solvent under reduced pressure the crude product was purified by flash column chromatography, eluting with HE/EA = 5/1; yield: 896 mg (82%), inseparable mixture of diastereomers, 73 mg (8%) of starting material recovered. The allylic alcohol obtained (896 mg, 3.28 mmol) was subsequently added to a suspension of NaH (197 mg, 4.92 mmol) in dry THF (20 ml) under argon at 0 °C. After stirring for 30 min, TBAI (242 mg, 0.66 mmol) and BnBr (1.2 ml, 10.10 mmol) were added and the reaction mixture was heated to reflux for 16 h. The reaction was quenched by adding saturated ammonium chloride solution (20 ml). After separation of the layers, the aqueous phase was extracted three times with ether (20 ml) and the combined organic extracts were dried over anhydrous MgSO₄ and filtered. After removal of the solvent under reduced pressure the crude product was purified by flash column chromatography, eluting with HE/EA = 9/1; yield: 900 mg (76%), inseparable mixture of diastereomers, 125 mg (14%) of starting material recovered. The benzylated olefin (900 mg, 2.49 mmol) in dry DCM (50 ml) was subsequently subjected to ozonolysis, quenching with PPh₃ (784 mg, 2.99 mmol) according to method A. Purification by silica gel chromatography was performed using HE/EA = 9/1 as eluent; yield: 643 mg (71%), (main diastereomer), 71 mg (8%), (minor diastereomer), (dr = 9/1). For spectroscopic data of compounds see Ref. 105.

(R)-tert-butyl 4-((1R,2S,E)-1-(benzyloxy)-5-ethoxy-2-fluoro-5-oxopent-3-en-1-yl)-2,2-dimethyloxazolidine-3-carboxylate (13)

A solution of compound **12** (40 mg, 0.11 mmol) and the catalyst (7 mg, 0.02 mmol) in MTBE (3 ml) was stirred for 10 min at room temperature and subsequently cooled to -20 °C, whereupon NFSI (35 mg, 0.11 mmol) was added. The reaction mixture was stirred at -20 to -15 °C for 40 h and subsequently diluted with dry DCM (3 ml) and treated with ethyl 2-(triphenylphosphoranylidene)acetate (77 mg, 0.22 mmol). The reaction mixture was allowed to warm to room temperature and stirred for 6 h, whereupon the solvents were removed under reduced pressure. The crude residue was subjected to purification by silica gel chromatography using HE/EA = 9/1 as eluent; yield: 10 mg (20%). ^{1}H NMR (CDCl$_3$, 600 MHz, 25°C): δ = 1.30 (t, ^3J = 7.1 Hz, 3 H, CH$_2$-CH$_3$), 1.47, 1.49, 1.52, 1.54, 1.62 (5 s, 5 CCH$_3$, cf), 3.89-4.04 (m, 2 H, 6-H, 7a-H), 4.11-4.25 (m, 2 H, 5-H, 7b-H), 4.21 (q, ^3J = 7.1 Hz, 2 H, O-CH$_2$-CH$_3$), 4.59-4.72 (m, cf, 2 H, O-CH$_2$-Ph), 5.14-5.28 (m, 1 H, 4-H), 6.18 (d, $^3J_{2,3}$ = 15.4 Hz, 1 H, 2-H), 6.95-7.13 (m, 1 H, 3-H), 7.27-7.36 (m, 5 H, CH-Ar), ^{13}C NMR (CDCl$_3$, 150 MHz, 25°C): δ = 14.4 (CH$_3$), 23.4, 25.2, 26.7, 27.1, 28.5, 28.7 (cf, 5 CCH$_3$), 57.2, 57.6 (cf, 6-C), 60.7 (O-CH$_2$-CH$_3$), 64.1, 64.5 (cf, 7-C), 74.3, 74.5 (cf, O-CH$_2$-Ph), 78.9 (cf, d, $^2J_{4-F,5}$ = 20.3 Hz, 5-C), 80.9, 81.1 (cf, Cq-Boc), 92.6 (cf, d, $^1J_{4-F,4}$ = 180.0 Hz, 4-C), 122.8 (cf, d, $^3J_{4-F,2}$ = 8.9 Hz, 2-C), 128.1, 128.3, 128.6 (CH-Ar), 137.7 (Cq-Ar), 141.5, 141.7 (cf, 3-C), 165.8 (1-C), ^{19}F NMR (CDCl$_3$, 600 MHz, 25°C): δ = -192.93, -189.59 (cf); HRMS (ESI): calcd. for C$_{24}$H$_{34}$FNNaO$_6$ [M + Na]$^+$ 474.2268, found 474.2277.

(4S,5R)-3-(2-fluoroacetyl)-4-methyl-5-phenyloxazolidin-2-one (14)

A stirred solution of (4S,5R)-4-methyl-5-phenyloxazolidin-2-one (14.52 g, 81.94 mmol) in 300 ml of dry tetrahydrofurane (THF) under an argon atmosphere was cooled to -78°C and BuLi 1.6 M in hexanes (53.8 ml, 86.08

mmol) was added dropwise. The reaction mixture was stirred at -78°C for 30 min, and then flouroacetyl chloride (6.3 ml, 90.10 mmol) was added dropwise. After additional stirring for 10 min the reaction mixture was warmed to 0°C and quenched by the addition of water. After separation of the phases the aqueous layer was extracted three times with Et$_2$O (100 ml). The combined organic extracts were washed with brine, dried over anhydrous MgSO$_4$ and filtered. After removal of the solvent under reduced pressure the crude product was purified by flash column chromatography eluting with hexanes (HE)/EA = 9/1; yield: 11.08 g (57%). $[\alpha]^D_{20}$ = -23.5° (6.9, CH$_2$Cl$_2$); m.p. 87-89°C; ^{1}H NMR (CDCl$_3$, 400 MHz, 25°C): δ = 0.96 (d, $^3J_{H,H}$ = 6.72 Hz, 3 H, CH$_3$), 4.80 (dq, $^3J_{H,H}$ = 7.08 Hz, 1 H, 4-H), 5.43 (dd, $^2J_{H,H}$ = 16.55 Hz, $^2J_{F,H}$ = 47.45 Hz, 1 H, 2a'-H), 5.47 (dd, 1 H, 2b'-H), 5.79 (d, 1 H, 5-H), 7.27-7.32 (m, 2 H, CH-Ar), 7.36-7.47 (m, 3 H, CH-Ar); ^{13}C NMR (CDCl$_3$, 100 MHz, 25°C): δ = 14.54 (CH$_3$), 54.45 (4-C), 79.03, 81.09 (2'-C), 80.67 (5-C), 125.63, 128.85, 129.10 (CH-Ar), 132.70 (Cq-Ar), 152.99 (2-C), 167.15, 167.36 (1'-C), ^{19}F NMR (CDCl$_3$, 600 MHz, 25°C): δ = -229.89; HRMS (ESI): calcd. for C$_{12}$H$_{12}$FNNaO$_3$ [M + Na]$^+$ 260.0699, found 260.0709.

(S)-tert-butyl-4-((1S,2R)-2-fluoro-1-hydroxy-3-((4S,5R)-4-methyl-2-oxo-5-phenyloxazolidin-3-yl)-3-oxopropyl)-2,2-dimethyloxazolidine-3-carboxylate (15)

(4S,5R)-3-(2-fluoroacetyl)-4-methyl-5-phenyloxazolidin-2-one (1.35 g, 5.67 mmol) in dry DCM (40 ml) was treated according to method D with TiCl$_4$ (670 µl, 6.11 mmol), freshly distilled TMEDA (2.6 ml, 17.45 mmol) and the aldehyde (1 g, 4.36 mmol). Purification by silica gel chromatography was performed using HE/EA = 3/1 as eluent; yield: (main diastereomers, could not be separated at this stage) 1.24 g, (61%), (dr = 4/1), (minor diastereomers) 142 mg, (7%), (dr = 1/1). ^{1}H NMR (CDCl$_3$, 400 MHz, 25°C), (main diastereomer): δ = 0.98 (d, $^3J_{4',CH3}$ = 6.8 Hz, 3 H, CH$_3$), 1.50, 1.52, 1.61 (3s, 15 H, 5 CCH$_3$), 4.03 (dd, $^3J_{4,5a}$ = 6.0 Hz, $^2J_{5a,5b}$ = 9.4 Hz, 1 H, 5a-H), 4.09-4.26 (m, 2 H, 3-H, 5b-H), 4.28-4.45 (m, 1 H, 4-H), 4.64 (d, $^3J_{3-OH,3}$ = 7.1 Hz, 1 H, OH), 4.77 (dq, $^3J_{4',5'}$ = 6.7 Hz, $^3J_{4',CH3}$ = 6.8 Hz, 1 H, 4'-H), 5.78 (d, $^3J_{4',5'}$ = 6.7 Hz, 1 H, 5'-H), 5.96 (d, $^2J_{2-F,2}$ = 48.8 Hz, 1 H, 2-H), 7.27-7.46 (m, 5 H, CH-Ar), ^{13}C NMR (CDCl$_3$, 100 MHz, 25°C): δ = 14.2 (4'-CH$_3$), 24.3, 27.3, 28.4 (5 CCH$_3$), 55.5 (4'-C), 58.3 (4-C), 64.4 (5-C), 73.9 (d, $^2J_{2-F,3}$ = 18.9 Hz, 3-C), 80.3 (5'-C), 81.8 (Cq-Boc), 89.4 (d, $^1J_{2-F,2}$ = 185.8 Hz, 2-C), 94.7 (Cq-Isoprop), 125.6, 128.8, 129.0 (CH-Ar), 132.6 (Cq-Ar), 152.9 (2'-C), 166.5 (d, $^2J_{2-F,1}$ = 24.1 Hz, 1-C), ^{19}F NMR (CDCl$_3$, 565 MHz, 25°C): δ = -211.02, HRMS (ESI): calcd. for C$_{23}$H$_{31}$FN$_2$NaO$_7$ [M + Na]$^+$ 489.2013, found 489.2012.

(S)-tert-butyl-4-((1R,2S)-2-fluoro-1-hydroxy-3-((4S,5R)-4-methyl-2-oxo-5-phenyloxazolidin-3-yl)-3-oxopropyl)-2,2-dimethyloxazolidine-3-carboxylate (16)

(4S,5R)-3-(2-fluoroacetyl)-4-methyl-5-phenyloxazolidin-2-one (1.35 g, 5.67 mmol) in dry DCM (40 ml) was treated according to method D with TiCl$_4$ (670 μl, 6.11 mmol), freshly distilled TMEDA (2.6 ml, 17.45 mmol) and the aldehyde (1 g, 4.36 mmol). Purification by silica gel chromatography was performed using HE/EA = 2/1 as eluent; yield: (main diastereomers) 1.18 g, (58%), (dr = 17/1), (minor diastereomers) 163 mg, (8%), (dr = 1/1). [α]$^D_{20}$ = +4.6° (7.1, CH$_2$Cl$_2$); m.p. 87-90 °C; ^{1}H NMR (CDCl$_3$, 400 MHz, 25°C): δ = 0.97 (d, $^3J_{4',CH3}$ = 6.6 Hz, 3 H, CH$_3$), 1.51, 1.64 (2s, 15 H, 5 CCH$_3$), 3.97-4.23 (m, 2 H, 3-H, 5a-H), 4.28-4.37 (m, 2 H, 4-H, 5b-H), 4.78 (dq, $^3J_{4',5'}$ = 7.1 Hz, $^3J_{4',CH3}$ = 6.6 Hz, 1

H, 4'-H), 4.90 (d, $^3J_{3\text{-OH},3}$ = 9.2 Hz, 1 H, OH), 5.76 (d, $^3J_{4',5'}$ = 7.1 Hz, 1 H, 5'-H), 6.16 (d, $^2J_{2\text{-F},2}$ = 47.8 Hz, 1 H, 2-H), 7.27-7.31 (m, 2 H, CH-Ar), 7.34-7.46 (m, 3 H, CH-Ar), ^{13}C NMR (CDCl$_3$, 100 MHz, 25°C): δ = 14.2 (4'-CH$_3$), 23.9, 26.5, 28.4 (5 CCH$_3$), 55.5 (4'-C), 60.1 (4-C), 64.6 (5-C), 72.9 (d, $^2J_{2\text{-F},3}$ = 20.2 Hz, 3-C), 80.3 (5'-C), 81.6 (Cq-Boc), 89.1 (d, $^1J_{2\text{-F},2}$ = 183.6 Hz, 2-C), 94.7 (Cq-Isoprop), 125.6, 128.8, 129.0 (CH-Ar), 132.7 (Cq-Ar), 152.9 (2'-C), 167.3 (1-C), ^{19}F NMR (CDCl$_3$, 565 MHz, 25°C): δ = -210.99; HRMS (ESI): calcd. for C$_{23}$H$_{31}$FN$_2$NaO$_7$ [M + Na]$^+$ 489.2013, found 489.2017.

4R,5R)-tert-butyl-4-((1R,2S)-2-fluoro-1-hydroxy-3-((4S,5R)-4-methyl-2-oxo-5-phenyloxazolidin-3-yl)-3-oxopropyl)-2,2,5-trimethyloxazolidine-3-carboxylate (17)

(4S,5R)-3-(2-fluoroacetyl)-4-methyl-5-phenyloxazolidin-2-one (1.92 g, 8.09 mmol) in dry DCM (60 ml) was treated according to method D with TiCl$_4$ (950 μl, 8.66 mmol), freshly distilled TMEDA (3.8 ml, 25.18 mmol) and the aldehyde (1.516 g, 6.23 mmol). Purification by silica gel chromatography was performed using HE/EA = 4/1 as eluent; yield: (main diastereomers) 1.20 g, (40%), (dr = 5/1), (minor diastereomers) 120 mg, (4%), (dr = 1/1). [α]$^D_{20}$ = -67.7° (8.2, CH$_2$Cl$_2$); m.p. 164-167 °C; ^{1}H NMR (CDCl$_3$, 400 MHz, 25°C): δ = 0.98 (d, $^3J_{4',4'\text{-CH3}}$ = 6.8 Hz, 3 H, 4'-CH$_3$), 1.41 (d, $^3J_{5,6}$ = 6.3 Hz, 3 H, 6-CH$_3$), 1.49, 1.52, 1.64 (3s, 15 H, 5 CCH$_3$), 4.12-4.27 (m, 2 H, 3-H, 4-H), 4.39 (dq, $^3J_{4,5}$ = 3.1 Hz, $^3J_{5,6}$ = 6.3 Hz, 1 H, 5-H), 4.78 (dq, $^3J_{4',4'\text{-CH3}}$ = 6.8 Hz, $^3J_{4',5'}$ = 7.1 Hz, 1 H, 4'-H), 4.91 (brs, 1 H, OH), 5.79 (d, $^3J_{4',5'}$ = 7.1 Hz, 1 H, 5'-H), 5.92 (dd, $^3J_{2,3}$ = 1.3 Hz, $^2J_{2\text{-F},2}$ = 48.5 Hz, 1 H, 2-H), 7.26-7.31 (m, 2 H, CH-Ar), 7.35-7.46 (m, 3 H, CH-Ar), ^{13}C NMR (CDCl$_3$, 100 MHz, 25°C): δ = 14.6 (4'-CH$_3$), 21.8 (6-C), 28.1, 28.7, 29.4 (5 CCH$_3$), 55.9 (4'-C), 64.7 (4-C), 73.2 (5-C), 74.9 (3-C), 80.8 (5'-C),

82.2 (Cq-Boc), 89.3 (d, $^1J_{2\text{-}F,2}$ = 184.3 Hz, 2-C), 95.0 (Cq-Isoprop), 126.0, 129.2 (CH-Ar), 133.0 (Cq-Ar), 153.4 (2'-C), 166.3 (d, $^2J_{2\text{-}F,1}$ = 24.1 Hz, 1-C), ^{19}F NMR (CDCl$_3$, 565 MHz, 25°C): δ = -209.80; HRMS (ESI): calcd. for C$_{24}$H$_{33}$FN$_2$NaO$_7$ [M + Na]$^+$ 503.2170, found 503.2176.

(4S,5S)-tert-butyl 4-((1R,2S)-2-fluoro-1-hydroxy-3-((4S,5R)-4-methyl-2-oxo-5-phenyloxazolidin-3-yl)-3-oxopropyl)-2,2,5-trimethyloxazolidine-3-carboxylate (18)

(4S,5R)-3-(2-fluoroacetyl)-4-methyl-5-phenyloxazolidin-2-one (0.76 g, 3.20 mmol) in dry DCM (25 ml) was treated according to method D with TiCl$_4$ (400 μl, 3.65 mmol), freshly distilled TMEDA (1.5 ml, 9.94 mmol) and the aldehyde (0.60 g, 2.47 mmol). Purification by silica gel chromatography was performed using HE/EA = 4/1 as eluent; yield: (main diastereomers) 624 mg, (53%), (dr = 16/1), (minor diastereomers) 27 mg, (2%), (dr = 2/1). [α]$^D_{20}$ = -4.8° (9.3, CH$_2$Cl$_2$); m.p. 70-73 °C; ^{1}H NMR (CDCl$_3$, 400 MHz, 25°C): δ = 0.97 (d, $^3J_{4',4'\text{-}CH3}$ = 6.6 Hz, 3 H, 4'-CH$_3$), 1.40 (d, $^3J_{5,6}$ = 6.3 Hz, 3 H, 6-CH$_3$), 1.50, 1.51, 1.61 (3s, 9 H, 5 CCH$_3$), 3.91 (dd, $^3J_{3,4}$ = 2.3 Hz, $^3J_{4,5}$ = 6.8 Hz, 1 H, H-4), 4.01-4.17 (m, 1 H, 3-H), 4.33-4.44 (m, 1 H, 5-H), 4.79 (dq, $^3J_{4',4'\text{-}CH3}$ = 6.6 Hz, $^3J_{4',5'}$ = 7.1 Hz, 1 H, 4'-H), 5.78 (d, $^3J_{4',5'}$ = 7.1 Hz, 1 H, 5'-H), 6.08 (d, $^2J_{2\text{-}F,2}$ = 48.3 Hz, 1 H, 2-H), 6.15 (brs, 1 H, OH), 7.27-7.31 (m, 2 H, CH-Ar), 7.36-7.45 (m, 3 H, CH-Ar), ^{13}C NMR (CDCl$_3$, 100 MHz, 25°C): δ = 14.2 (4'-CH$_3$), 19.5 (6-C), 26.4, 28.5, 28.6 (5 CCH$_3$), 55.7 (4'-C), 67.9 (4-C), 70.7 (d, $^2J_{2\text{-}F,3}$ = 18.5 Hz, 3-C), 71.9 (5-C), 80.5 (5'-C), 81.8 (Cq-Boc), 89.2 (d, $^1J_{2\text{-}F,2}$ = 187.9 Hz, 2-C), 94.7 (Cq-Isoprop), 125.7, 128.9, 129.1 (CH-Ar), 132.8 (Cq-Ar), 153.2 (2'-C), 167.2 (d, $^2J_{2\text{-}F,1}$ = 24.7 Hz, 1-C),^{19}F NMR (CDCl$_3$, 565 MHz, 25°C): δ = -210.78; HRMS (ESI): calcd. for C$_{24}$H$_{33}$FN$_2$NaO$_7$ [M + Na]$^+$ 503.2170, found 503.2160.

(4S,5R)-tert-butyl-4-((1R,2S)-2-fluoro-1-hydroxy-3-((4S,5R)-4-methyl-2-oxo-5-phenyloxazolidin-3-yl)-3-oxopropyl)-2,2,5-trimethyloxazolidine-3-carboxylate (19)

(4S,5R)-3-(2-fluoroacetyl)-4-methyl-5-phenyloxazolidin-2-one (512 mg, 2.16 mmol) in dry DCM (20 ml) was treated according to method D with TiCl$_4$ (260 µl, 2.37 mmol), freshly distilled TMEDA (1 ml, 6.63 mmol) and the aldehyde (404 mg, 1.66 mmol). Purification by silica gel chromatography was performed using HE/EA = 4/1 as eluent; yield: (main diastereomers) 374 mg, (47%), (dr = 20/1). $[\alpha]^D_{20}$ = -7.9° (6.4, CH$_2$Cl$_2$); m.p. 80-83 °C; ^{1}H NMR (CDCl$_3$, 400 MHz, 25°C): δ = 0.98 (d, $^3J_{4',4'-CH3}$ = 6.6 Hz, 3 H, 4'-CH$_3$), 1.46 (d, $^3J_{5,6}$ = 6.6 Hz, 3 H, 6-CH$_3$), 1.51, 1.56, 1.66 (3s, 9 H, 5 CCH$_3$), 4.09-4.22 (m, 2 H, 3-H, 4-H), 4.31-4.40 (m, 1 H, 5-H), 4.62 (brs, 1 H, OH), 4.77 (dq, $^3J_{4',4'-CH3}$ = 6.6 Hz, $^3J_{4',5'}$ = 6.8 Hz, 1 H, 4'-H), 5.76 (d, $^3J_{4',5'}$ = 6.8 Hz, 1 H, 5'-H), 6.10 (d, $^2J_{2-F,2}$ = 47.0 Hz, 1 H, 2-H), 7.27-7.31 (m, 2 H, CH-Ar), 7.35-7.45 (m, 3 H, CH-Ar), ^{13}C NMR (CDCl$_3$, 150 MHz, 25°C): δ = 14.2 (4'-CH$_3$), 15.0 (6-C), 24.6, 26.7, 28.4 (5 CCH$_3$), 55.7 (4'-C), 62.8 (4-C), 70.1 (d, $^2J_{2-F,3}$ = 19.4 Hz, 3-C), 72.0 (5-C), 80.2 (5'-C), 81.4 (Cq-Boc), 89.9 (d, $^1J_{2-F,2}$ = 184.3 Hz, 2-C), 93.4 (Cq-Isoprop), 125.6, 128.8, 129.0 (CH-Ar), 132.7 (Cq-Ar), 152.59 (2'-C), 154.46 (C=O-Boc), 167.3 (d, $^2J_{2-F,1}$ = 24.5 Hz, 1-C), ^{19}F NMR (CDCl$_3$, 565 MHz, 25°C): δ = -208.66; HRMS (ESI): calcd. for C$_{24}$H$_{33}$FN$_2$NaO$_7$ [M + Na]$^+$ 503.2170, found 503.2171.

(2S,3R,4S)-methyl-4-((tert-butoxycarbonyl)amino)-2-fluoro-3,5-dihydroxypentanoate (20)

Fluorohydrin **15** (788 mg, 1.69 mmol) in dry MeOH (20 ml) was cooled to -40°C and treated according to method E with NaOMe (18 mg, 0.34 mmol) and then with acidic ion exchange resin for 4 h. Purification by silica gel chromatography was performed using HE/EA = 1/1 as eluent; yield: (main diastereomer) 165 mg, (35%), (minor diastereomer) 33 mg, (7%). $[\alpha]^D_{20}$ = -6.1° (15.0, CH_2Cl_2); ^{1}H NMR ($CDCl_3$, 400 MHz, 25°C): δ = 1.43 (s, 9 H, 3 CH_3), 3.24 (brs, 1 H, 5-OH), 3.72-3.90 (m, 3 H, 4-H, 5a-H, 5b-H), 3.82 (s, 3 H, OCH_3), 4.30 (m, 1 H, 3-H), 5.06 (dd, $^3J_{2,3}$ = 3.0 Hz, $^2J_{2-F,2}$ = 48.0 Hz, 1 H, 2-H), 5.36 (d, $^3J_{4-NH,4}$ = 8.6 Hz, 1 H, NH), ^{13}C NMR ($CDCl_3$, 100 MHz, 25°C): δ = 28.7, 28.7 (3 CH_3), 53.1 (OCH_3), 53.7 (4-C), 63.3 (5-C), 70.7 (d, $^2J_{2-F,3}$ = 19.1 Hz, 3-C), 80.7 (Cq-Boc), 89.6 (d, $^1J_{2-F,2}$ = 188.9 Hz, 2-C), 157.0 (C=O-Boc), 168.6 (d, $^2J_{2-F,1}$ = 24.6 Hz, 1-C), ^{19}F NMR ($CDCl_3$, 565 MHz, 25°C): δ = -206.97; HRMS (ESI): calcd. for $C_{11}H_{20}FNNaO_6$ $[M + Na]^+$ 304.1172, found 304.1177.

(2S,3R,4R)-methyl-4-((tert-butoxycarbonyl)amino)-2-fluoro-3,5-dihydroxypentanoate (21)

Fluorohydrin **16** (486 mg, 1.04 mmol) in dry MeOH (10 ml) was cooled to -25°C and treated according to method E with NaOMe (23 mg, 0.42 mmol) and then with acidic ion exchange resin for 16h. Purification by silica gel chromatography was performed using HE/EA = 1/2 as eluent; yield: 121 mg (41%). $[\alpha]^D_{20}$ = -17.2° (4.7, CH_2Cl_2); m.p. 123-125 °C; ^{1}H NMR ($CDCl_3$, 400 MHz, 25°C): δ = 1.45 (s, 9 H, 3 CCH_3), 3.78-3.88 (m, 2 H, 4-H, 5a-H), 3.85 (s, 3 H, OCH_3), 4.06 (dd, $^3J_{4,5b}$ = 3.0 Hz, $^2J_{5a,5b}$ = 10.9 Hz, 1 H, 5b-H), 4.22 (m, 1 H, 3-H), 5.11 (dd, $^3J_{2,3}$ = 1.8 Hz, $^2J_{2-F,2}$ = 47.5 Hz, 1 H, 2-H), 5.29 (brs, 1 H, NH),

^{13}C NMR (CDCl$_3$, 100 MHz, 25°C): δ = 28.3 (3 CCH$_3$), 52.5 (4-C), 52.7 (OCH$_3$), 62.4 (5-C), 71.6 (d, $^2J_{2\text{-}F,3}$ = 19.8 Hz, 3-C), 80.4 (Cq-Boc), 89.0 (d, $^1J_{2\text{-}F,2}$ = 189.4 Hz, 2-C), 156.0 (C=O-Boc), 168.9 (d, $^2J_{2\text{-}F,1}$ = 25.5 Hz, 1-C), ^{19}F NMR (CDCl$_3$, 565 MHz, 25°C): δ = -208.49; HRMS (ESI): calcd. for C$_{11}$H$_{20}$FNNaO$_6$ [M + Na]$^+$ 304.1172, found 304.1178.

(2S,3R,4S,5R)-methyl-4-((tert-butoxycarbonyl)amino)-2-fluoro-3,5-dihydroxyhexanoate (22)

Fluorohydrin **17** (200 mg, 0.42 mmol) in dry MeOH (10 ml) was cooled to -30°C and treated according to method E with NaOMe (11 mg, 0.20 mmol) and then with acidic ion exchange resin for 2h. Purification by silica gel chromatography was performed using T/EA = 2/1 as eluent; yield: 70 mg, (57%). [α]$^D_{20}$ = -15.0° (13.9, CH$_2$Cl$_2$); ^{1}H NMR (CDCl$_3$, 400 MHz, 25°C): δ = 1.23 (d, $^3J_{5,6}$ = 6.2 Hz, 3 H, 6-CH$_3$), 1.44 (s, 9 H, 3 CCH$_3$), 2.71 (brs, 1 H, 5-OH), 3.42 (brs, 1 H, 3-OH), 3.71 (m, 1 H, 4-H), 3.83 (s, 3 H, OCH$_3$), 4.14 (m, 1 H, 5-H), 4.29 (m, 1 H, 3-H), 5.05 (dd, $^2J_{2\text{-}F,2}$ = 47.8 Hz, $^3J_{2,3}$ = 3.0 Hz, 1 H, 2-H), 5.24 (d, $^3J_{4\text{-}NH,4}$ = 9.60 Hz, 1 H, NH), ^{13}C NMR (CDCl$_3$, 100 MHz, 25°C): δ = 20.6 (6-C), 28.5 (3 CCH$_3$), 52.9 (OCH$_3$), 55.8 (4-C), 69.6 (5-C), 73.4 (d, $^2J_{2\text{-}F,3}$ = 19.1 Hz, 3-C), 80.1 (Cq-Boc), 89.3 (d, $^1J_{2\text{-}F,2}$ = 188.9 Hz, 2-C), 157.3 (C=O-Boc), 168.3 (d, $^2J_{2\text{-}F,1}$ = 23.8 Hz, 1-C), ^{19}F NMR (CDCl$_3$, 565 MHz, 25°C): δ = -207.49; HRMS (ESI): calcd. for C$_{12}$H$_{22}$FNNaO$_6$ [M + Na]$^+$ 318.1329, found 318.1336.

(2S,3R,4R,5S)-methyl-4-((tert-butoxycarbonyl)amino)-2-fluoro-3,5-dihydroxyhexanoate (23)

Fluorohydrin **18** (202 mg, 0.42 mmol) in dry MeOH (10 ml) was cooled to -30°C and treated according to method E with NaOMe (9 mg, 0.17 mmol) and then with acidic ion exchange resin for 2.5 h. Purification by silica gel chromatography was performed using toluene (T)/EA = 2/1 as eluent; yield: 70 mg, (56%). $[\alpha]^D_{20}$ = -10.8° (7.1, CH_2Cl_2); m.p. 138-140 °C; ^1H NMR (MeOD, 600 MHz, 25°C): δ = 1.15 (d, cf, $^3J_{5,6}$ = 6.5 Hz, 3 H, 6-CH_3), 1.45 (s, 9 H, 3 CCH_3), 3.62 (dd, $^3J_{4,5}$ = 1.5 Hz, $^3J_{3,4}$ = 10.4 Hz, 1 H, 4-H), 3.80 (s, 3 H, OCH_3), 4.04 (ddd, $^3J_{2,3}$ = 1.2 Hz, $^3J_{3,4}$ = 10.4 Hz, $^3J_{2-F,3}$ = 28.9 Hz, 1 H, 3-H), 4.20 (dq, $^3J_{4,5}$ = 1.5 Hz, $^3J_{5,6}$ = 6.5 Hz, 1 H, 5-H), 5.06 (dd, cf, $^3J_{2,3}$ = 1.2 Hz, $^2J_{2-F,2}$ = 47.8 Hz, 1 H, 2-H), ^{13}C NMR (MeOD, 150 MHz, 25°C): δ = 20.4 (cf, 6-C), 28.7 (cf, 3 CCH_3), 52.7 (OCH_3), 56.4 (d, cf, $^3J_{2-F,4}$ = 2.7 Hz, 4-C), 65.7 (cf, 5-C), 71.9 (d, $^2J_{2-F,3}$ = 19.3 Hz, 3-C), 80.4 (Cq-Boc), 90.3 (d, $^1J_{2-F,2}$ = 188.5 Hz, 2-C), 158.2 (C=O-Boc), 170.9 (d, cf, $^2J_{2-F,1}$ = 25.2 Hz, 1-C), ^{19}F NMR (MeOD, 565 MHz, 25°C): δ = -211.85 (cf); HRMS (ESI): calcd. for $C_{12}H_{22}FNNaO_6$ [M + Na]$^+$ 318.1329, found 318.1327.

4-acetamido-1,3-di-O-acetyl-2,4-dideoxy-2-fluoro-D-xylose (24)

Ester **20** (88 mg, 0.31 mmol) in dry THF (7 ml) was treated according to method F with DIBAL (1.9 ml, 1.9 mmol). The crude amino sugar was then dissolved in dry DCM (5 ml) and treated with Ac_2O (60 μl, 0.63 mmol) for 1 h at room temperature. After evaporation of the solvent, purification by silica gel chromatography was performed using HE/EA = 1/4 as eluent; yield: 49 mg, (mixture of anomers: α/β = 1/7), (56%). ^{1}H NMR (CDCl$_3$, 600 MHz, 25°C): (α-anomer) δ = 1.93 (s, 3 H, NHAc), 2.13, 2.15 (2s, 6 H, 2 OAc), 3.50 (dd, $^2J_{5a,5b}$ = 11.4 Hz, $^3J_{4,5a}$ = 11.3 Hz, 1 H, 5a-H), 3.91 (ddd, ^4J = 1.5 Hz, $^3J_{4,5b}$ = 5.5 Hz, $^2J_{5a,5b}$ = 11.4 Hz, 1 H, 5b-H), 4.18 (m, 1 H, 4-H), 4.63 (ddd, $^3J_{1,2}$ = 3.9 Hz, $^3J_{2,3}$ = 9.3 Hz, $^2J_{2\text{-}F,2}$ = 48.5 Hz, 1 H, 2-H), 5.28 (ddd, $^3J_{2,3}$ = 9.3 Hz, $^3J_{3,4}$ = 10.4 Hz, $^3J_{2\text{-}F,3}$ = 11.3 Hz, 1 H, 3-H), 5.93 (d, $^3J_{4\text{-}NH,4}$ = 7.7 Hz, 1 H, NH), 6.32 (dd, $^3J_{2\text{-}F,1}$ = 1.3 Hz, $^3J_{1,2}$ = 3.9 Hz, 1 H, 1-H), (β-anomer) δ = 1.95 (s, 3 H, NHAc), 2.12, 2.14 (2s, 6 H, 2 OAc), 3.38 (dd, $^3J_{4,5a}$ = 8.4 Hz, $^2J_{5a,5b}$ = 11.8 Hz, 1 H, 5a-H), 4.13 (dd, $^3J_{4,5b}$ = 4.7 Hz, $^2J_{5a,5b}$ = 11.8 Hz, 1 H, 5b-H), 4.18 (m, 1 H, 4-H), 4.43 (ddd, $^3J_{1,2}$ = 6.2 Hz, $^3J_{2,3}$ = 7.3 Hz, $^2J_{2\text{-}F,2}$ = 48.7 Hz, 1 H, 2-H), 5.08 (ddd, $^3J_{2,3}$ = 7.3 Hz, $^3J_{3,4}$ = 8.2 Hz, $^3J_{2\text{-}F,3}$ = 12.6 Hz, 1 H, 3-H), 5.78 (dd, $^3J_{2\text{-}F,1}$ = 5.9 Hz, $^3J_{1,2}$ = 6.2 Hz, 1 H, 1-H), 6.08 (d, $^3J_{4\text{-}NH,4}$ = 8.5 Hz, 1 H, NH), ^{13}C NMR (CDCl$_3$, 150 MHz, 25°C): (α-anomer) δ = 20.8, 20.9 (2 OAc), 23.1 (NHAc), 49.5 (d, $^3J_{2\text{-}F,4}$ = 5.8 Hz, 4-C), 62.0 (5-C), 70.3 (d, $^2J_{2\text{-}F,3}$ = 18.9 Hz, 3-C), 86.2 (d, $^2J_{2\text{-}F,2}$ = 193.0 Hz, 2-C), 89.0 (d, $^2J_{2\text{-}F,1}$ = 22.7 Hz, 1-C), 168.9, 172.0 (2 CO-OAc), 170.3 (CO-NHAc), (β-anomer) δ = 20.8, 20.8 (2 OAc), 23.1 (NHAc), 48.4 (d, $^3J_{2\text{-}F,4}$ = 4.4 Hz, 4-C), 63.6 (5-C), 70.8 (d, $^2J_{2\text{-}F,3}$ = 21.1 Hz, 3-C), 86.9 (d, $^1J_{2\text{-}F,2}$ = 186.5 Hz, 2-C), 91.4 (d, $^2J_{2\text{-}F,1}$ = 26.0 Hz, 1-C), 169.0, 171.0 (2 CO-OAc), 170.3 (CO-NHAc) ^{19}F NMR (CDCl$_3$, 565 MHz, 25°C): δ = -201.08, -197.06; HRMS (ESI): calcd. for C$_{11}$H$_{16}$FNNaO$_6$ [M + Na]$^+$ 300.0859, found 300.0845.

4-acetamido-1-O-acetyl-2,4-dideoxy-2-fluoro-D-arabinose (25)

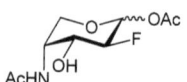

Ester **21** (94 mg, 0.33 mmol) in dry THF (7 ml) was treated according to method F with DIBAL (2 ml, 2 mmol). Purification by silica gel chromatography was performed using EA/MeOH = 99/1 as eluent; yield: 52 mg, (mixture of anomers:

α/β = 1/1), (66%). ^{1}H NMR (MeOD, 400 MHz, 25°C): (α-anomer) δ = 2.01 (s, 3 H, NHAc), 2.13 (s, 3 H, OAc), 3.66 (ddd, $^2J_{5a,5b}$ = 12.3 Hz, $^3J_{4,5a}$ = 3.9 Hz, ^4J = 1.5 Hz, 1 H, 5a-H), 3.92 (dd, $^2J_{5a,5b}$ = 12.3 Hz, $^3J_{4,5b}$ = 2.8 Hz, 1 H, 5b-H), 4.19 (ddd, $^3J_{2-F,3}$ = 10.7 Hz, $^3J_{2,3}$ = 8.7 Hz, $^3J_{3,4}$ = 4.6 Hz, 1 H, 3-H), 4.33-4.38 (m, 1 H, 4-H), 4.67 (ddd, $^3J_{1,2}$ = 3.4 Hz, $^3J_{2,3}$ = 8.7 Hz, $^2J_{2-F,2}$ = 48.2 Hz, 1 H, 2-H), 6.21 (dd, $^3J_{1,2}$ = 3.4 Hz, $^3J_{2-F,1}$ = 5.1 Hz, 1 H, 1-H), (β-anomer) δ = 2.00 (s, 3 H, NHAc), 2.11 (s, 3 H, OAc), 3.62 (dd, $^2J_{5a,5b}$ = 11.6 Hz, $^3J_{4,5a}$ = 3.0 Hz, 1 H, 5a-H), 3.87 (ddd, ^4J = 0.8 Hz , $^3J_{4,5b}$ = 5.7 Hz, $^2J_{5a,5b}$ = 11.6 Hz, 1 H, 5b-H), 4.05 (ddd, $^3J_{2-F,3}$ = 11.8 Hz, $^3J_{2,3}$ = 6.8 Hz, $^3J_{3,4}$ = 4.6 Hz, 1 H, 3-H), 4.25-4.30 (m, 1 H, 4-H), 4.51 (ddd, $^3J_{1,2}$ = 5.3 Hz, $^3J_{2,3}$ = 6.8 Hz, $^2J_{2-F,2}$ = 48.1 Hz, 1 H, 2-H), 5.74 (dd, $^3J_{1,2}$ = 5.3 Hz, $^3J_{2-F,1}$ = 7.1 Hz, 1 H, 1-H), ^{13}C NMR (MeOD, 100 MHz, 25°C): (α-anomer) δ = 21.1 (OAc), 22.9 (NHAc), 51.3 (d, $^3J_{2-F,4}$ = 6.6 Hz, 4-C), 64.8 (5-C), 67.9 (d, $^2J_{2-F,3}$ = 20.5 Hz, 3-C), 89.2 (d, $^1J_{2-F,2}$ = 186.8 Hz, 2-C), 91.5 (d, $^2J_{2-F,1}$ = 20.7 Hz, 1-C), 171.3 (CO-OAc), 174.1 (CO-NHAc), (β-anomer) δ = 21.1 (OAc), 23.0 (NHAc), 50.0 (d, $^3J_{2-F,4}$ = 4.5 Hz, 4-C), 64.3 (5-C), 69.6 (d, $^2J_{2-F,3}$ = 22.2 Hz, 3-C), 90.5 (d, $^1J_{2-F,2}$ = 179.4 Hz, 2-C), 93.4 (d, $^2J_{2-F,1}$ = 29.0 Hz, 1-C), 171.3 (CO-OAc), 174.4 (CO-NHAc), ^{19}F NMR (MeOD, 565 MHz, 25°C): δ = -201.19, -208.69; HRMS (ESI): calcd. for $C_9H_{14}FNNaO_5$ [M + Na]$^+$ 258.0754, found 258.0751.

4-acetamido-1,3-di-O-acetyl-2,4,6-trideoxy-2-fluoro-D-idose (26)

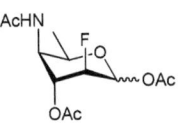

Ester **22** (66 mg, 0.22 mmol) in dry THF (6 ml) was treated according to method F with DIBAL (1.4 ml, 1.4 mmol). The crude amino sugar was then dissolved in dry DCM (5 ml) and treated with Ac$_2$O (40 μl, 0.42 mmol) for 1 h at room temperature. After evaporation of the solvent, purification by silica gel chromatography was performed using HE/EA = 1/2 as eluent; yield: 33 mg, (mixture of anomers: α/β = 2/3), (53%). ^{1}H NMR (CDCl$_3$, 600 MHz, 25°C): (α-anomer) δ = 1.19 (d, $^3J_{5,6}$ = 6.5 Hz, 3 H, 6-CH$_3$), 2.04 (s, 3 H, NHAc), 2.11, 2.12 (2s, 6 H, 2 OAc), 4.15 (m, 1 H, 4-H), 4.43 (dq, $^3J_{4,5}$ = 2.1 Hz, $^3J_{5,6}$ = 6.5 Hz, 1 H, 5-H), 4.45-4.47, 4.52-4.54 (m, 1 H, 2-H), 5.01-5.04 (m, 1 H, 3-H), 5.92 (d, $^3J_{4\text{-}NH,4}$ = 10.5 Hz, 1 H, NH), 6.15 (d, $^3J_{2\text{-}F,1}$ = 11.8 Hz, 1 H, 1-H), (β-anomer) δ = 1.24 (d, $^3J_{5,6}$ = 6.3 Hz, 3 H, 6-CH$_3$), 2.03 (s, 3 H, NHAc), 2.13, 2.19 (2s, 6 H, 2 OAc), 4.15 (m, 1 H, 4-H), 4.23 (dq, $^3J_{4,5}$ = 2.3 Hz, $^3J_{5,6}$ = 6.3 Hz, 1 H, 5-H), 4.50-4.52, 4.58-4.60 (m, 1 H, 2-H), 5.19 (m, 1 H, 3-H), 5.91 (d, $^3J_{2\text{-}F,1}$ = 22.8 Hz, 1 H, 1-H), 6.01 (d, $^3J_{4\text{-}NH,4}$ = 9.2 Hz, 1 H, NH),^{13}C NMR (CDCl$_3$, 150 MHz, 25°C): (α-anomer) δ = 16.5 (6-C), 20.8, 21.0 (2 OAc), 23.3 (NHAc), 47.8 (4-C), 64.5 (5-C), 67.1 (d, $^2J_{2\text{-}F,3}$ = 27.7 Hz, 3-C), 83.2 (d, $^1J_{2\text{-}F,2}$ = 171.6 Hz, 2-C), 90.5 (d, $^2J_{2\text{-}F,1}$ = 35.3 Hz, 1-C), 168.6, 169.2 (2 CO-OAc), 170.2 (CO-NHAc), (β-anomer) δ = 16.5 (6-C), 20.9, 21.0 (2 OAc), 23.3 (NHAc), 47.9 (4-C), 68.8 (d, $^2J_{2\text{-}F,3}$ = 25.8 Hz, 3-C), 71.4 (5-C), 84.4 (d, $^1J_{2\text{-}F,2}$ = 184.9 Hz, 2-C), 90.4 (d, $^2J_{2\text{-}F,1}$ = 15.2 Hz, 1-C), 168.8, 168.9 (2 CO-OAc), 170.0 (CO-NHAc), ^{19}F NMR (CDCl$_3$, 565 MHz, 25°C): δ = -190.05, -208.98; HRMS (ESI): calcd. for C$_{12}$H$_{18}$FNNaO$_6$ [M + Na]$^+$ 314.1016, found 314.0998.

4-acetamido-1-O-acetyl-2,4,6-trideoxy-2-fluoro-L-galactose (27)

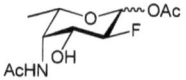

Ester **23** (69 mg, 0.23 mmol) in dry THF (5 ml) was treated according to method F with DIBAL (1.6 ml, 1.6 mmol). After the final work up, the crude product was stirred for 1 h at room temperature in methanolic solution. Purification by silica gel chromatography was performed using EA/HE = 4/1 as eluent; yield: 27 mg, (mixture of anomers: α/β = 2/3), (47%). ^{1}H NMR (MeOD, 600 MHz, 25°C): (α-anomer) δ = 1.08 (d, $^3J_{5,6}$ = 6.4 Hz, 3 H, 6-CH$_3$), 2.05 (s, 3 H, NHAc), 2.13 (s,

3 H, OAc), 4.21 (ddd, $^3J_{3,4}$ = 4.8 Hz, $^3J_{2,3}$ = 10.2 Hz, $^3J_{2\text{-}F,3}$ = 12.7 Hz, 1 H, 3-H), 4.22 (dq, $^3J_{4,5}$ = 2.0 Hz, $^3J_{5,6}$ = 6.4 Hz, 1 H, 5-H), 4.40 (m, 1 H, 4-H), 4.66 (ddd, $^3J_{1,2}$ = 4.2 Hz, $^3J_{2,3}$ = 10.2 Hz, $^2J_{2\text{-}F,2}$ = 48.6 Hz, 1 H, 2-H), 6.28 (d, $^3J_{1,2}$ = 4.2 Hz, 1 H, 1-H), (β-anomer) δ = 1.13 (d, $^3J_{5,6}$ = 6.4 Hz, 3 H, 6-CH$_3$), 2.05 (s, 3 H, NHAc), 2.12 (s, 3 H, OAc), 3.92 (dq, $^3J_{4,5}$ = 1.7 Hz, $^3J_{5,6}$ = 6.4 Hz, 1 H, 5-H), 4.03 (ddd, $^3J_{3,4}$ = 5.1 Hz, $^3J_{2,3}$ = 9.7 Hz, $^3J_{2\text{-}F,3}$ = 14.9 Hz, 1 H, 3-H), 4.33 (m, 1 H, 4-H), 4.45 (ddd, $^3J_{1,2}$ = 8.0 Hz, $^3J_{2,3}$ = 9.7 Hz, $^2J_{2\text{-}F,2}$ = 51.6 Hz, 1 H, 2-H), 5.64 (dd, $^3J_{2\text{-}F,1}$ = 4.4 Hz, $^3J_{1,2}$ = 8.0 Hz, 1 H, 1-H), ^{13}C NMR (MeOD, 150 MHz, 25°C): (α-anomer) δ = 16.7 (6-C), 20.7 (OAc), 22.5 (NHAc), 55.3 (d, $^3J_{2\text{-}F,4}$ = 8.4 Hz, 4-C), 68.4 (d, $^2J_{2\text{-}F,3}$ = 17.9 Hz, 3-C), 69.1 (5-C), 88.8 (d, $^1J_{2\text{-}F,2}$ = 186.7 Hz, 2-C), 90.8 (1-C), 171.0 (CO-OAc), 174.7 (CO-NHAc), (β-anomer) δ = 16.6 (6-C), 20.7 (OAc), 22.4 (NHAc), 55.1 (d, $^3J_{2\text{-}F,4}$ = 8.7 Hz, 4-C), 71.7 (d, $^2J_{2\text{-}F,3}$ = 17.6 Hz, 3-C), 72.3 (5-C), 91.4 (d, $^1J_{2\text{-}F,2}$ = 159.9 Hz, 2-C), 93.5 (d, $^2J_{2\text{-}F,1}$ = 25.0 Hz, 1-C), 170.9 (CO-OAc), 174.7 (CO-NHAc), ^{19}F NMR (CDCl$_3$, 565 MHz, 25°C): δ = -208.92, -210.42; HRMS (ESI): calcd. for C$_{10}$H$_{16}$FNNaO$_5$ [M + Na]$^+$ 272.0912, found 272.0908.

4-acetamido-2,4-dideoxy-2-fluoro-D-xylose (28)

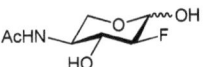

Acetylated sugar 24 (32 mg, 0.12 mmol) in dry MeOH (3 ml) was treated according to method C; yield: 22 mg, (mixture of anomers: α/β = 2/3), (100%). [α]$^D_{20}$ = -45.9° (8.5, H$_2$O); ^{1}H NMR (D$_2$O, 600 MHz, 25°C): (α-anomer) δ = 1.94 (s, 3 H, NHAc), 3.59 (ddd, ^4J = 1.8 Hz, $^3J_{4,5a}$ = 6.2 Hz, $^2J_{5a,5b}$ = 11.4 Hz, 1 H, 5a-H), 3.62 (dd, $^3J_{4,5b}$ = 10.1 Hz, $^2J_{5a,5b}$ = 11.4 Hz, 1 H, 5b-H), 3.86 (ddd, $^3J_{4,5a}$ = 6.2 Hz, $^3J_{3,4}$ = 9.9 Hz, $^3J_{4,5b}$ = 10.1 Hz, 1 H, 4-H), 3.94 (ddd, $^3J_{2,3}$ = 9.0 Hz, $^3J_{3,4}$ = 9.9 Hz, $^3J_{2\text{-}F,3}$ = 12.4 Hz, 1 H, 3-H), 4.41 (ddd, $^3J_{1,2}$ = 3.7 Hz, $^3J_{2,3}$ = 9.0 Hz, $^2J_{2\text{-}F,2}$ = 48.9 Hz, 1 H, 2-H), 5.35 (d, $^3J_{1,2}$ = 3.7 Hz, 1 H, 1-H), (β-anomer) δ = 1.94 (s, 3

H, NHAc), 3.26 (dd, $^3J_{4,5a}$ = 10.5 Hz, $^2J_{5a,5b}$ = 11.4 Hz, 1 H, 5a-H), 3.78 (ddd, $^3J_{2,3}$ = 8.8 Hz, $^3J_{3,4}$ = 10.0 Hz, $^3J_{2\text{-}F,3}$ = 14.3 Hz, 1 H, 3-H), 3.83 (dd, $^3J_{4,5b}$ = 5.2 Hz, $^2J_{5a,5b}$ = 11.4 Hz, 1 H, 5b-H), 3.87 (ddd, $^3J_{4,5b}$ = 5.2 Hz, $^3J_{3,4}$ = 10.0 Hz, $^3J_{4,5a}$ = 10.5 Hz, 1 H, 4-H), 4.08 (ddd, $^3J_{1,2}$ = 7.8 Hz, $^3J_{2,3}$ = 8.8 Hz, $^2J_{2\text{-}F,2}$ = 51.0 Hz, 1 H, 2-H), 4.75 (dd, $^3J_{2\text{-}F,1}$ = 2.8 Hz, $^3J_{1,2}$ = 7.8 Hz, 1 H, 1-H), ^{13}C NMR (D$_2$O, 150 MHz, 25°C): (α-anomer) δ = 21.9 (NHAc), 50.5 (d, $^3J_{2\text{-}F,4}$ = 7.3 Hz, 4-C), 59.1 (5-C), 68.4 (d, $^2J_{2\text{-}F,3}$ = 18.7 Hz, 3-C), 90.0 (d, $^2J_{2\text{-}F,1}$ = 21.1 Hz, 1-C), 90.6 (d, $^1J_{2\text{-}F,2}$ = 184.9 Hz, 2-C), 174.6 (CO-NHAc), (β-anomer) δ = 21.9 (NHAc), 50.7 (d, $^3J_{2\text{-}F,4}$ = 7.9 Hz, 4-C), 63.3 (5-C), 71.5 (d, $^2J_{2\text{-}F,3}$ = 18.4 Hz, 3-C), 93.3 (d, $^1J_{2\text{-}F,2}$ = 182.6 Hz, 2-C), 94.2 (d, $^2J_{2\text{-}F,1}$ = 23.3 Hz, 1-C), 174.7 (CO-NHAc), ^{19}F NMR (CDCl$_3$, 565 MHz, 25°C): δ = -199.71, -199.04; HRMS (ESI): calcd. for C$_7$H$_{12}$FNNaO$_4$ [M + Na]$^+$ 216.0648, found 216.0636.

4-acetamido-2,4-dideoxy-2-fluoro-D-arabinose (29)

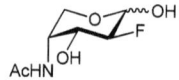

Acetylated sugar **25** (52 mg, 0.22 mmol) in dry MeOH (5 ml) was treated according to method C. Purification by silica gel chromatography was performed using DCM/MeOH = 9/1 as eluent; yield: 38 mg, (mixture of anomers: α/β = 1/2), (89%). [α]$^D_{28}$ = -66.8° (14.0, H$_2$O); ^{1}H NMR (D$_2$O, 600 MHz, 25°C): (α-anomer) δ = 2.06 (s, 3 H, NHAc), 3.42 (ddd, ^4J = 1.8 Hz, $^3J_{4,5a}$ = 4.5 Hz, $^2J_{5a,5b}$ = 12.2 Hz, 1 H, 5a-H), 4.06 (dd, $^3J_{4,5b}$ = 3.0 Hz, $^2J_{5a,5b}$ = 12.2 Hz, 1 H, 5b-H), 4.29 (ddd, $^3J_{3,4}$ = 4.4 Hz, $^3J_{2,3}$ = 8.4 Hz, $^3J_{2\text{-}F,3}$ = 10.4 Hz, 1 H, 3-H), 4.35 (m, 1 H, 4-H), 4.62 (ddd, $^3J_{1,2}$ = 3.1 Hz, $^3J_{2,3}$ = 8.4 Hz, $^2J_{2\text{-}F,2}$ = 48.2 Hz, 1 H, 2-H), 5.39 (dd, $^3J_{1,2}$ = 3.1 Hz, $^3J_{2\text{-}F,1}$ = 6.3 Hz, 1 H, 1-H), (β-anomer) δ = 2.08 (s, 3 H, NHAc), 3.75 (dd, $^3J_{4,5a}$ = 2.2 Hz, $^2J_{5a,5b}$ = 12.7 Hz, 1 H, 5a-H), 3.87 (ddd, ^4J = 1.5 Hz, $^3J_{4,5b}$ = 2.5 Hz, $^2J_{5a,5b}$ = 12.7 Hz, 1 H, 5b-H), 4.15 (ddd, $^3J_{3,4}$ = 4.9 Hz, $^3J_{2,3}$ = 9.2 Hz, $^3J_{2\text{-}F,3}$ = 14.0 Hz, 1 H, 3-H), 4.32 (ddd, $^3J_{1,2}$ = 7.2 Hz, $^3J_{2,3}$ = 9.2 Hz,

$^2J_{2-F,2}$ = 50.5 Hz, 1 H, 2-H), 4.34 (m, 1 H, 4-H), 4.84 (dd, $^3J_{2-F,1}$ = 3.9 Hz, $^3J_{1,2}$ = 7.2 Hz, 1 H, 1-H), ^{13}C NMR (D$_2$O, 150 MHz, 25°C): (α-anomer) δ = 21.9 (CH$_3$), 49.1 (d, $^3J_{2-F,4}$ = 4.8 Hz, 4-C), 60.9 (5-C), 65.9 (d, $^2J_{2-F,3}$ = 21.8 Hz, 3-C), 88.7 (d, $^1J_{2-F,2}$ = 185.1 Hz, 2-C), 90.3 (d, $^2J_{2-F,1}$ = 20.1 Hz, 1-C), 174.6 (CO), (β-anomer) δ = 21.9 (CH$_3$), 50.1 (d, $^3J_{2-F,4}$ = 8.2 Hz, 4-C), 64.5 (5-C), 69.4 (d, $^2J_{2-F,3}$ = 17.2 Hz, 3-C), 91.9 (d, $^1J_{2-F,2}$ = 177.1 Hz, 2-C), 94.2 (d, $^2J_{2-F,1}$ = 24.1 Hz, 1-C), 174.7 (CO), ^{19}F NMR (CDCl$_3$, 565 MHz, 25°C): δ = -207.10, -203.18; HRMS (ESI): calcd. for C$_7$H$_{12}$FNNaO$_4$ [M + Na]$^+$ 216.0648, found 216.0644.

4-acetamido-2,4,6-trideoxy-2-fluoro-D-idose (30)

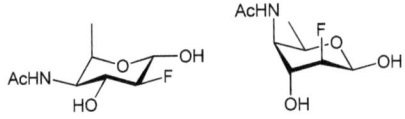

Acetylated sugar **26** (19 mg, 0.07 mmol) in dry MeOH (2 ml) was treated according to method C; yield: 13 mg, (mixture of anomers: α/β = 1/2), (97%). [α]$^D_{20}$ = +18.3° (5.9, H$_2$O); ^{1}H NMR (D$_2$O, 600 MHz, 25°C): (α-anomer) δ = 1.19 (d, $^3J_{5,6}$ = 6.9 Hz, 3 H, 6-CH$_3$), 2.02 (s, 3 H, NHAc), 3.95 (dd, $^3J_{4,5}$ = 4.4 Hz, $^3J_{3,4}$ = 7.3 Hz, 1 H, 4-H), 4.02 (ddd, $^3J_{2,3}$ = 6.4 Hz, $^3J_{3,4}$ = 7.3 Hz, $^3J_{2-F,3}$ = 12.1 Hz, 1 H, 3-H), 4.29 (ddd, $^3J_{1,2}$ = 5.0 Hz, $^3J_{2,3}$ = 6.4 Hz, $^2J_{2-F,2}$ = 48.7 Hz, 1 H, 2-H), 4.40 (dq, $^3J_{4,5}$ = 4.4 Hz, $^3J_{5,6}$ = 6.9 Hz, 1 H, 5-H), 5.17 (dd, $^3J_{1,2}$ = 5.0 Hz, $^3J_{2-F,1}$ = 8.1 Hz, 1 H, 1-H), (β-anomer) δ = 1.17 (d, $^3J_{5,6}$ = 6.6 Hz, 3 H, 6-CH$_3$), 2.04 (s, 3 H, NHAc), 3.82 (dddd, ^4J = 0.6 Hz, $^4J_{2-F,4}$ = 0.9 Hz, $^3J_{4,5}$ = 2.4 Hz, $^3J_{3,4}$ = 3.0 Hz, 1 H, 4-H), 4.14 (ddd, $^3J_{3,4}$ = 3.0 Hz, $^3J_{2,3}$ = 3.4 Hz, $^3J_{2-F,3}$ = 6.7 Hz, 1 H, 3-H), 4.24 (dq, $^3J_{4,5}$ = 2.4 Hz, $^3J_{5,6}$ = 6.6 Hz, 1 H, 5-H), 4.45 (dddd, $^3J_{1,2}$ = 1.0 Hz, ^4J = 1.0 Hz, $^3J_{2,3}$ = 3.4 Hz, $^2J_{2-F,2}$ = 46.0 Hz, 1 H, 2-H), 5.09 (ddd, ^4J = 0.5 Hz, $^3J_{1,2}$ = 1.0 Hz, $^3J_{2-F,1}$ = 23.2 Hz, 1 H, 1-H), ^{13}C NMR (D$_2$O, 150 MHz, 25°C): (α-anomer) δ = 13.9 (6-C), 21.7 (NHAc), 52.0 (d, $^3J_{2-F,4}$ = 3.9 Hz, 4-C), 65.5 (5-C), 67.2 (d, $^2J_{2-F,3}$ = 21.5 Hz, 3-C), 90.3 (d, $^2J_{2-F,1}$ = 28.6 Hz, 1-C), 90.9 (d, $^1J_{2-F,2}$ =

177.1 Hz, 2-C), 174.4 (CO-NHAc), (β-anomer) δ = 15.8 (6-C), 21.7 (NHAc), 50.4 (4-C), 67.3 (d, $^2J_{2-F,3}$ = 24.0 Hz, 3-C), 68.9 (5-C), 87.8 (d, $^1J_{2-F,2}$ = 179.9 Hz, 2-C), 91.2 (d, $^2J_{2-F,1}$ = 15.9 Hz, 1-C), 174.3 (CO-NHAc), ^{19}F NMR (D$_2$O, 565 MHz, 25°C): δ = -194.82, -211.86; HRMS (ESI): calcd. for C$_8$H$_{14}$FNNaO$_4$ [M + Na]$^+$ 230.0805, found 230.0796.

4-acetamido-2,4,6-trideoxy-2-fluoro-D-galactose (31)

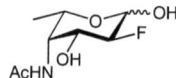

Acetylated sugar **27** (20 mg, 0.08 mmol) in dry MeOH (ml) was treated according to method C; yield: 16 mg, (mixture of anomers: α/β = 1/2), (96%). [α]$^D_{20}$ = -66.7° (4.7, H$_2$O); ^{1}H NMR (D$_2$O, 600 MHz, 25°C): (α-anomer) δ = 1.11 (d, $^3J_{5,6}$ = 6.5 Hz, 3 H, 6-CH$_3$), 2.11 (s, 3 H, NHAc), 4.30 (ddd, $^3J_{3,4}$ = 4.7 Hz, $^3J_{2,3}$ = 10.3 Hz, $^3J_{2-F,3}$ = 13.2 Hz, 1 H, 3-H), 4.35 (m, 1 H, 4-H), 4.42 (dq, $^3J_{4,5}$ = 1.7 Hz, $^3J_{5,6}$ = 6.5 Hz, 1 H, 5-H), 4.58 (ddd, $^3J_{1,2}$ = 4.1 Hz, $^3J_{2,3}$ = 10.3 Hz, $^2J_{2-F,2}$ = 49.0 Hz, 1 H, 2-H), 5.44 (d, $^3J_{1,2}$ = 4.1 Hz, 1 H, 1-H), (β-anomer) δ = 1.16 (d, $^3J_{5,6}$ = 6.4 Hz, 3 H, 6-CH$_3$), 2.12 (s, 3 H, NHAc), 3.99 (dq, $^3J_{4,5}$ = 1.6 Hz, $^3J_{5,6}$ = 6.4 Hz, 1 H, 5-H), 4.14 (ddd, $^3J_{3,4}$ = 4.9 Hz, $^3J_{2,3}$ = 9.8 Hz, $^3J_{2-F,3}$ = 14.8 Hz, 1 H, 3-H), 4.24 (ddd, $^3J_{1,2}$ = 7.8 Hz, $^3J_{2,3}$ = 9.8 Hz, $^2J_{2-F,2}$ = 51.0 Hz, 1 H, 2-H), 4.31 (m, 1 H, 4-H), 4.86 (dd, $^3J_{2-F,1}$ = 3.6 Hz, $^3J_{1,2}$ = 7.8 Hz, 1 H, 1-H), ^{13}C NMR (D$_2$O, 150 MHz, 25°C): (α-anomer) δ = 15.4 (6-C), 21.8 (NHAc), 54.3 (d, $^3J_{2-F,4}$ = 8.3 Hz, 4-C), 65.0 (5-C), 66.7 (d, $^2J_{2-F,3}$ = 17.9 Hz, 3-C), 88.8 (d, $^1J_{2-F,2}$ = 183.3 Hz, 2-C), 89.8 (d, $^2J_{2-F,1}$ = 21.1 Hz, 1-C), 175.5 (CO-NHAc), (β-anomer) δ = 15.4 (6-C), 21.8 (NHAc), 54.0 (d, $^3J_{2-F,4}$ = 8.8 Hz, 4-C), 70.1 (d, $^2J_{2-F,3}$ = 17.5 Hz, 3-C), 70.1 (5-C), 92.2 (d, $^1J_{2-F,2}$ = 180.6 Hz, 2-C), 93.9 (d, $^2J_{2-F,1}$ = 23.6 Hz, 1-C), 175.5 (CO-NHAc), ^{19}F NMR (D$_2$O, 565 MHz, 25°C): δ = -206.13, -206.34; HRMS (ESI): calcd. for C$_8$H$_{14}$FNNaO$_4$ [M + Na]$^+$ 230.0805, found 230.0804.

(S)-tert-butyl 4-((1R,2S)-2-fluoro-1-hydroxy-3-((4S,5R)-4-methyl-2-oxo-5-phenyloxazolidin-3-yl)-3-oxopropyl)-2,2-dimethylthiazolidine-3-carboxylate (32)

(4S,5R)-3-(2-fluoroacetyl)-4-methyl-5-phenyloxazolidin-2-one (663 mg, 2.79 mmol) in dry DCM (20 ml) was treated according to method D with TiCl$_4$ (330 µl, 3.01 mmol), freshly distilled TMEDA (1.3 ml, 8.59 mmol) and the aldehyde (527 mg, 2.15 mmol). Purification by silica gel chromatography was performed using EA/HE = 1/3 as eluent; yield: (main diastereomer) 509 mg, (49%), (minor diastereomers) 117 mg, (11%), 43 mg, (4%), (dr = 7/5). $[\alpha]_D^{20}$ = +29.9° (9.7, CH$_2$Cl$_2$); m.p. 206-209 °C; ^{1}H-NMR (400 MHz, CDCl$_3$, 25 °C): δ = 0.96 (d, $^3J_{4',4'-CH3}$ = 6.4 Hz, 3 H, 4'-CH$_3$), 1.50, 1.79, 1.83 (3 s, 15 H, 5 CCH$_3$), 3.23 (brs, 2 H, 5a-H, 5b-H), 4.31 (d, $^2J_{2-F,3}$ = 23.7 Hz, 1 H, 3-H), 4.72 (brs, 1 H, 4-H), 4.78 (dq, $^3J_{4',4'-CH3}$ = 6.4 Hz, $^3J_{4',5'}$ = 7.0 Hz, 1 H, 4'-H), 5.78 (d, $^3J_{4',5'}$ = 7.0 Hz, 1 H, 5'-H), 6.33 (d, $^2J_{2-F,2}$ = 46.0 Hz, 1 H, 2-H), 7.27-7.32 (m, 2 H, CH-Ar), 7.34-7.46 (m, 3 H, CH-Ar), ^{13}C-NMR (100 MHz, CDCl$_3$, 25 °C): δ = 14.3 (4'-CH$_3$), 28.5, 29.8 (5 CCH$_3$), 29.6 (5-C), 55.6 (4'-C), 65.5 (4-C), 73.0 (3-C), 80.3 (5'-C), 81.6 (Cq-Boc), 89.8 (d, $^1J_{2-F,2}$ = 183.4 Hz, 2-C), 125.7, 128.9, 129.0 (CH-Ar), 132.8 (Cq-Ar), 152.8 (2'-C), 167.6 ($^2J_{2-F,1}$ = 24.9 Hz, 1-C), ^{19}F NMR (CDCl$_3$, 565 MHz, 25°C): (cf) δ = -207.96, -210.69; HRMS (ESI): calcd. for C$_{23}$H$_{31}$FN$_2$NaO$_6$S [M + Na]$^+$ 505.1785, found 505.1770.

(R)-tert-butyl 4-((1R,2S)-2-fluoro-1-hydroxy-3-((4S,5R)-4-methyl-2-oxo-5-phenyloxazolidin-3-yl)-3-oxopropyl)-2,2-dimethylthiazolidine-3-carboxylate (33)

(4S,5R)-3-(2-fluoroacetyl)-4-methyl-5-phenyloxazolidin-2-one (1.63 g, 6.89 mmol) in dry DCM (50 ml) was treated according to method D with $TiCl_4$ (810 μl, 7.42 mmol), freshly distilled TMEDA (3.2 ml, 21.20 mmol) and the aldehyde (1.3 g, 5.30 mmol). Purification by silica gel chromatography was performed using EA/HE = 1/3 as eluent; yield: (main diastereomer) 1.51 g, (59%), (minor diastereomers) 0.23 g, (9%), (dr = 3/2/1). $[\alpha]_D^{20}$ = -112.3° (8.0, CH_2Cl_2); m.p. 177-180 °C; ^{1}H-NMR (400 MHz, $CDCl_3$, 25 °C): δ = 0.99 (d, $^3J_{4',4'-CH3}$ = 6.6 Hz, 3 H, 4'-CH_3), 1.50, 1.79, 1.81 (3 s, 15 H, 5 CCH_3), 2.95 (d, $^2J_{5a,5b}$ = 12.4 Hz, 1 H, 5a-H), 3.25 (dd, $^3J_{4,5b}$ = 5.8 Hz, $^2J_{5a,5b}$ = 12.4 Hz, 1 H, 5b-H), 4.17-4.46 (m, 2 H, 3-H, 3-OH), 4.77 (dq, $^3J_{4',4'-CH3}$ = 6.6 Hz, $^3J_{4',5'}$ = 7.0 Hz, 1 H, 4'-H), 4.81-4.95 (m, 1 H, 4-H), 5.80 (d, $^3J_{4',5'}$ = 7.0 Hz, 1 H, 5'-H), 5.98 (d, $^2J_{2-F,2}$ = 48.7 Hz, 1 H, 2-H), 7.27-7.31 (m, 2 H, CH-Ar), 7.36-7.46 (m, 3 H, CH-Ar), ^{13}C-NMR (100 MHz, $CDCl_3$, 25 °C): δ = 14.3 (4'-CH_3), 28.5, 30.8 (5 CCH_3), 29.2 (5-C), 55.7 (4'-C), 64.2 (4-C), 74.0 (3-C), 80.5 (5'-C), 82.0 (Cq-Boc), 89.8 (d, $^1J_{2-F,2}$ = 184.6 Hz, 2-C), 125.7, 128.9, 129.1 (CH-Ar), 132.7 (Cq-Ar), 153.2 (2'-C), 166.8 ($^2J_{2-F,1}$ = 23.8 Hz, 1-C), ^{19}F NMR ($CDCl_3$, 565 MHz, 25°C): (cf) δ = -211.69, -212.29; HRMS (ESI): calcd. for $C_{23}H_{31}FN_2NaO_6S$ [M + Na]$^+$ 505.1785, found 505.1793.

(S)-tert-butyl 4-((1R,2R)-2-fluoro-1,3-dihydroxypropyl)-2,2-dimethylthiazolidine-3-carboxylate (34)

Fluoro-hydrin **32** (420 mg, 0.87 mmol) was treated according to method G with LiBH$_4$ (28 mg, 1.31 mmol) and dry MeOH (50 µl, 1.23 mmol) in dry THF (25 ml). Purification by silica gel chromatography was performed using EA/HE = 2/3 as eluent; yield: 254 mg (94%) as a white crystalline solid. $[\alpha]_D^{20}$ = +26.1° (6.1, CH$_2$Cl$_2$); m.p. 113-115 °C; ^{1}H-NMR (600 MHz, CDCl$_3$, 25 °C): δ = 1.48, 1.78, 1.79 (3 s, 15 H, 5 CCH$_3$), 3.16-3.23 (m, 2 H, 5a-H, 5b-H), 2.53 (brs, 1 H, 1-OH), 3.58 (brs, 1 H, 3-OH), 3.85-3.94 (m, 1 H, 1a-H), 3.99 (ddd, $^3J_{1b,2}$ = 5.2 Hz, $^2J_{1a,1b}$ = 12.5 Hz, $^3J_{2\text{-F},1b}$ = 23.2 Hz, 1 H, 1b-H), 4.10-4.19 (m, 1 H, 3-H), 4.56 (m, 1 H, 4-H), 4.63 (d, $^2J_{2\text{-F},2}$ = 48.0 Hz, 1 H, 2-H), ^{13}C-NMR (150 MHz, CDCl$_3$, 25 °C): δ = 28.5, 29.2 (5 CCH$_3$), 29.1 (5-C), 63.7 (d, $^2J_{2\text{-F},1}$ = 23.8 Hz, 1-C), 65.6 (4-C), 73.2 (d, $^2J_{2\text{-F},1}$ = 17.0 Hz, 3-C), 81.4 (Cq-Boc), 92.6 (d, $^1J_{2\text{-F},2}$ = 177.9 Hz, 2-C), ^{19}F NMR (CDCl$_3$, 565 MHz, 25°C): δ = -208.17; HRMS (ESI): calcd. for C$_{13}$H$_{24}$FNNaO$_4$S [M + Na]$^+$ 332.1308, found 332.1301.

(R)-tert-butyl 4-((1R,2R)-2-fluoro-1,3-dihydroxypropyl)-2,2-dimethylthiazolidine-3-carboxylate (35)

Fluoro-hydrin **33** (1.51 g, 3.13 mmol) was treated according to method G with LiBH$_4$ (102 mg, 4.69 mmol) and dry MeOH (190 µl, 4.68 mmol) in dry THF (100 ml). Purification by silica gel chromatography was performed using EA/HE = 2/3 as eluent; yield: 919 mg (95%) as a colorless oil. $[\alpha]_D^{20}$ = -54.3° (23.0, CH$_2$Cl$_2$); ^{1}H-NMR (600 MHz, CDCl$_3$, 25 °C): δ = 1.48, 1.76, 1.79 (3 s, 15

H, 5 CCH$_3$), 2.65 (d, $^2J_{5a,5b}$ = 12.5 Hz, 1 H, 5a-H), 2.80 (brs, 1 H, 1-OH), 3.22 (ddd, $^4J_{3,5b}$ = 1.8 Hz, $^3J_{4,5b}$ = 6.0 Hz, $^2J_{5a,5b}$ = 12.5 Hz, 1 H, 5b-H), 3.82-3.92 (m, 1 H, 1a-H), 3.96-4.08 (m, 2 H, 1b-H, 3-H), 4.31 (brs, 1 H, 3-OH), 4.50-4.54, 4.58-4.62 (m, 1 H, 2-H), 4.72-4.85 (m, 1 H, 4-H), ^{13}C-NMR (150 MHz, CDCl$_3$, 25 °C): δ = 28.5, 29.2, 30.8 (5 CCH$_3$), 29.7 (5-C), 63.0 (d, $^2J_{2-F,1}$ = 23.5 Hz, 1-C), 64.8 (4-C), 74.5 (3-C), 82.1 (Cq-Boc), 92.7 (d, $^1J_{2-F,2}$ = 177.7 Hz, 2-C), ^{19}F NMR (CDCl$_3$, 565 MHz, 25°C): δ = -210.13; HRMS (ESI): calcd. for C$_{13}$H$_{24}$FNNaO$_4$S [M + Na]$^+$ 332.1308, found 332.1310.

(2R,3R,4R)-4-((2,4-dinitrophenyl)amino)-5-((2,4-dinitrophenyl)thio)-2-fluoropentane-1,3-diol (36)

Compound **15** (321 mg, 1.04 mmol) was deprotected under acidic conditions, according to method H and subsequently dissolved in MeOH (12 ml) and treated with TEA (580 μl, 4.18 mmol), followed by Sanger's reagent (260 μl, 2.07 mmol). The resulting bright yellow solution was stirred at room temperature for 18 h and subsequently evaporated to dryness. Purification by silica gel chromatography was performed using EA/HE = 2/1 as eluent; yield: 403 mg (78%). ^{1}H-NMR (400 MHz, MeOD, 25 °C): δ = 3.62 (dd, $^3J_{1a,2}$ = 5.8 Hz, $^2J_{1a,1b}$ = 13.9 Hz, 1 H, 1a-H), 3.69 (ddd, $^3J_{4,5a}$ = 5.2 Hz, $^2J_{5a,5b}$ = 12.6 Hz, $^3J_{4-F,5a}$ = 23.6 Hz, 1 H, 5a-H), 3.71 (dd, $^3J_{1b,2}$ = 7.8 Hz, $^2J_{1a,1b}$ = 13.9 Hz, 1 H, 1b-H), 3.77 (ddd, $^3J_{4,5b}$ = 3.9 Hz, $^2J_{5a,5b}$ = 12.6 Hz, $^3J_{4-F,5b}$ = 22.7 Hz, 1 H, 5b-H), 4.19 (ddd, $^3J_{2,3}$ = 2.3 Hz, $^3J_{3,4}$ = 5.3 Hz, $^3J_{4-F,3}$ = 17.9 Hz, 1 H, 3-H), 4.51 (ddd, $^3J_{2,3}$ = 2.3 Hz, $^3J_{1a,2}$ = 5.8 Hz, $^3J_{1b,2}$ = 7.8 Hz, 1 H, 2-H), 4.52 (dddd, $^3J_{4,5b}$ = 3.9 Hz, $^3J_{4,5a}$ = 5.2 Hz, $^3J_{3,4}$ = 5.3 Hz, $^2J_{4-F,4}$ = 47.7 Hz, 1 H, 4-H), 7.20 (d, ^3J = 9.7 Hz, 1 H, CH-Ar),

7.99 (d, ^3J = 9.0 Hz, 1 H, CH-Ar), 8.18 (dd, ^4J = 2.7 Hz, ^3J = 9.7 Hz, 1 H, CH-Ar), 8.39 (dd, ^4J = 2.5 Hz, ^3J = 9.0 Hz, 1 H, CH-Ar), 8.80 (d, ^4J = 2.5 Hz, 1 H, CH-Ar), 8.91 (d, ^4J = 2.7 Hz, 1 H, CH-Ar), ^{13}C-NMR (100 MHz, MeOD, 25 °C): δ = 36.4 (1-C), 54.7 (d, $^3J_{4-F,2}$ = 5.6 Hz, 2-C), 62.2 (d, $^2J_{4-F,5}$ = 24.0 Hz, 5-C), 71.5 (d, $^2J_{4-F,3}$ = 20.9 Hz, 3-C), 95.9 (d, $^1J_{4-F,4}$ = 175.0 Hz, 4-C), 116.2, 122.0, 124.5, 128.2, 130.0, 131.0 (CH-Ar), 131.7, 137.5, 145.4, 145.8, 147.6, 149.1 (Cq-Ar), ^{19}F NMR (MeOD, 565 MHz, 25°C): δ = -201.53; HRMS (ESI): calcd. for C17H16FN5NaO10S [M + Na]$^+$ 524.0500, found 524.0488.

2-acetamido-2,4-dideoxy-4-fluoro-D-lyxose (40)

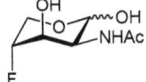

Thiazolidine **34** (102 mg, 0.33 mmol) was treated according to method H with TPP (2 mg, 0.003 mmol) in MTBE (5 ml) and quenched with PPh$_3$ (138 mg, 0.50 mmol); yield: 24 mg (22%), 67 mg (66%) of starting material recovered. Hydroxy-thiazolidine **38** (32 mg, 0.10 mmol) was then treated with 3 M HCl/EA = 1/1 (3 ml) followed by TEA (20 μl, 0.14 mmol) and Ac$_2$O (9 μl, 0.10 mmol) in dry MeCN (2 ml); yield: 11 mg (58%). [α]$_D^{20}$ = +13.5° (4.0, H$_2$O); ^{1}H-NMR (600 MHz, D$_2$O, 25 °C): (α/β = 3/1), (α-anomer) δ = 2.06 (s, 3 H, NHAc), 4.01 (ddd, $^3J_{4,5a}$ = 1.9 Hz, $^2J_{5a,5b}$ = 13.5 Hz, $^3J_{4-F,5a}$ = 32.9 Hz, 1 H, 5a-H), 4.03 (dd, $^3J_{2,3}$ = 3.4 Hz, $^3J_{1,2}$ = 7.7 Hz, 1 H, 2-H), 4.05 (dddd, ^4J = 1.2 Hz, $^3J_{4,5b}$ = 3.1 Hz, $^2J_{5a,5b}$ = 13.5 Hz, $^3J_{4-F,5b}$ = 14.9 Hz, 1 H, 5b-H), 4.22 (dddd, ^4J = 1.1 Hz, $^3J_{2,3}$ = 3.4 Hz, $^3J_{3,4}$ = 4.8 Hz, $^3J_{4-F,3}$ = 11.0 Hz, 1 H, 3-H), 4.65 (dddd, $^3J_{4,5a}$ = 1.9 Hz, $^3J_{4,5b}$ = 3.1 Hz, $^3J_{3,4}$ = 4.8 Hz, $^2J_{4-F,4}$ = 45.5 Hz, 1 H, 4-H), 4.95 (dd, $^3J_{1,2}$ = 7.7 Hz, $^5J_{4-F,1}$ = 1.5 Hz, 1 H, 1-H), (β-anomer) δ = 2.09 (s, 3 H, NHAc), 3.70 (ddd, $^3J_{4,5a}$ = 5.0 Hz, $^2J_{5a,5b}$ = 13.1 Hz, $^3J_{4-F,5a}$ = 11.1 Hz, 1 H, 5a-H), 4.12 (ddd, $^3J_{2,3}$ = 4.0 Hz, $^3J_{3,4}$ = 6.0 Hz, $^3J_{4-F,3}$ = 8.9 Hz, 1 H, 3-H), 4.24 (ddd, $^3J_{4,5b}$ = 2.9 Hz, $^2J_{5a,5b}$ = 13.1 Hz, $^3J_{4-F,5b}$ = 26.7 Hz, 1 H, 5b-H), 4.31 (dd, $^3J_{1,2}$ = 3.0 Hz, $^3J_{2,3}$ = 4.0 Hz, 1

H, 2-H), 4.71 (dddd, $^3J_{4,5b}$ = 2.9 Hz, $^3J_{4,5a}$ = 5.0 Hz, $^3J_{3,4}$ = 6.0 Hz, $^2J_{4-F,4}$ = 46.6 Hz, 1 H, 4-H), 5.11 (d, $^3J_{1,2}$ = 3.0 Hz, 1 H, 1-H), ^{13}C-NMR (150 MHz, D$_2$O, 25 °C): (α-anomer) δ = 21.9 (NHAc), 51.8 (2-C), 62.8 (d, $^2J_{4-F,5}$ = 20.7 Hz, 5-C), 66.9 (d, $^2J_{4-F,3}$ = 27.4 Hz, 3-C), 88.0 (d, $^1J_{4-F,4}$ = 174.5 Hz, 4-C), 92.3 (1-C), 174.3 (CO-NHAc), (β-anomer) δ = 21.9 (NHAc), 49.2 (2-C), 58.7 (d, $^2J_{4-F,5}$ = 23.1 Hz, 5-C), 67.3 (d, $^2J_{4-F,3}$ = 24.8 Hz, 3-C), 88.2 (d, $^1J_{4-F,4}$ = 175.0 Hz, 4-C), 91.9 (1-C), 174.7 (CO-NHAc), ^{19}F NMR (D$_2$O, 565 MHz, 25°C): δ = -194.67 (α-anomer), -198.74 (β-anomer); HRMS (ESI): calcd. for C$_7$H$_{12}$FNNaO$_4$ [M + Na]$^+$ 216.0648, found 216.0643.

2-acetamido-2,4-dideoxy-4-fluoro-D-xylose (41)

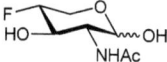

Thiazolidine **35** (400 mg, 1.29 mmol) was treated according to method H with TPP (8 mg, 0.01 mmol) in MTBE (10 ml) and quenched with PPh$_3$ (540 mg, 1.94 mmol); yield: 96 mg (23%), 264 mg (66%) of starting material recovered. Hydroxy-thiazolidine **39** (77 mg, 0.24 mmol) was then treated with 3 M HCl/EA = 1/1 (6 ml) followed by TEA (50 µl, 0.36 mmol) and Ac$_2$O (22 µl, 0.23 mmol) in dry MeCN (3 ml); yield: 27 mg (59%). [α]$_D^{20}$ = +13.5° (4.0, H$_2$O); ^{1}H-NMR (600 MHz, D$_2$O, 25 °C): mixture of anomers, α/β = 3/2, (α-anomer) δ = 2.05 (s, 3 H, NHAc), 3.88-3.96 (m, 3 H, 2-H, 5a-H, 5b-H), 4.00 (ddd, $^3J_{3,4}$ = 8.0 Hz, $^3J_{2,3}$ = 9.9 Hz, $^3J_{4-F,3}$ = 13.8 Hz, 1 H, 3-H), 4.56 (dddd, $^3J_{4,5a}$ = 6.1 Hz, $^3J_{3,4}$ = 8.0 Hz, $^3J_{4,5b}$ = 9.3 Hz, $^2J_{4-F,4}$ = 49.6 Hz, 1 H, 4-H), 5.17 (dd, $^3J_{1,2}$ = 3.3 Hz, $^5J_{4-F,1}$ = 3.3 Hz, 1 H, 1-H), (β-anomer) δ = 2.05 (s, 3 H, NHAc), 3.55 (ddd, $^3J_{4-F,5a}$ = 4.4 Hz, $^3J_{4,5a}$ = 9.9 Hz, $^2J_{5a,5b}$ = 11.7 Hz, 1 H, 5a-H), 3.73 (dd, $^3J_{1,2}$ = 8.0 Hz, $^3J_{2,3}$ = 10.2 Hz, 1 H, 2-H), 3.82 (ddd, $^3J_{2,3}$ = 10.2 Hz, $^3J_{3,4}$ = 8.3 Hz, $^3J_{4-F,3}$ = 14.8 Hz, 1 H, 3-H), 4.17 (ddd, $^3J_{4,5b}$ = 5.5 Hz, $^3J_{4-F,5b}$ = 2.7 Hz, $^2J_{5a,5b}$ = 11.7 Hz, 1 H, 5b-H), 4.58 (dddd, $^3J_{4,5b}$ = 5.5 Hz, $^3J_{3,4}$ = 8.3 Hz, $^3J_{4,5a}$ = 9.9 Hz, $^2J_{4-F,4}$ = 49.8 Hz, 1 H, 4-H),

4.71 (d, $^3J_{1,2}$ = 8.0 Hz, 1 H, 1-H), ^{13}C-NMR (150 MHz, D$_2$O, 25 °C): (α-anomer) δ = 21.8 (NHAc), 53.2 (d, $^3J_{4-F,2}$ = 7.5 Hz, 2-C), 58.9 (d, $^2J_{4-F,5}$ = 27.7 Hz, 5-C), 69.0 (d, $^2J_{4-F,3}$ = 19.2 Hz, 3-C), 89.3 (d, $^1J_{4-F,4}$ = 177.4 Hz, 4-C), 90.7 (d, $^4J_{4-F,1}$ = 1.2 Hz, 1-C), 174.5 (CO-NHAc), (β-anomer) δ = 22.1 (NHAc), 55.6 (d, $^3J_{4-F,2}$ = 8.7 Hz, 2-C), 62.2 (d, $^2J_{4-F,5}$ = 28.6 Hz, 5-C), 71.8 (d, $^2J_{4-F,3}$ = 18.8 Hz, 3-C), 89.2 (d, $^1J_{4-F,4}$ = 178.1 Hz, 4-C), 95.3 (d, $^4J_{4-F,1}$ = 1.1 Hz, 1-C), 174.7 (CO-NHAc), ^{19}F NMR (D$_2$O, 565 MHz, 25°C): δ = -199.86 (α-anomer), -197.00 (β-anomer); HRMS (ESI): calcd. for C$_7$H$_{12}$FNNaO$_4$ [M + Na]$^+$ 216.0648, found 216.0647.

(2R,3R,4R,5R,6S)-6-fluoro-5-hydroxy-7-((4S,5R)-4-methyl-2-oxo-5-phenyloxazolidin-3-yl)-7-oxoheptane-1,2,3,4-tetrayl tetraacetate (42)

(4S,5R)-3-(2-fluoroacetyl)-4-methyl-5-phenyloxazolidin-2-one (197 mg, 0.83 mmol) in dry DCM (10 ml) was treated according to method D with TiCl$_4$ (100 μl, 0.91 mmol), freshly distilled DIPEA (220 μl, 1.29 mmol) and the aldehyde (203 mg, 0.64 mmol). Purification by silica gel chromatography was performed using EA/HE = 1/1 as eluent; yield: 111 mg (31%), (main diastereomer, contaminated with minor amounts of auxilliary), 120 mg (34%), (minor diastereomers, ratio not determined). ^{1}H-NMR (400 MHz, CDCl$_3$, 25 °C): δ = 0.95 (d, $^3J_{4',4'-CH3}$ = 6.6 Hz, 3 H, 4'-CH$_3$), 2.06, 2.07, 2.13, 2.18 (4 s, 12 H, 4 OAc), 2.63 (d, $^3J_{3-OH,3}$ = 11.5 Hz, 1 H, 3-OH), 4.11 (dd, $^3J_{6,7a}$ = 5.2 Hz, $^2J_{7a,7b}$ = 12.5 Hz, 1 H, 7a-H), 4.25 (dd, $^3J_{6,7b}$ = 2.6 Hz, $^2J_{7a,7b}$ = 12.5 Hz, 1 H, 7b-H), 4.38 (dddd, $^3J_{2,3}$ = 1.5 Hz, $^3J_{3,4}$ = 6.1 Hz, $^3J_{3-OH,3}$ = 11.5 Hz, $^3J_{2-F,3}$ = 27.1 Hz, 1 H, 3-H), 4.75 (dq, $^3J_{4',4'-CH3}$ = 6.6 Hz, $^3J_{4',5'}$ = 7.1 Hz, 1 H, 4'-H), 5.20 (ddd, $^3J_{6,7b}$ = 2.6 Hz, $^3J_{6,7a}$ = 5.2 Hz, $^3J_{5,6}$ = 8.6 Hz, 1 H, 6-H), 5.45 (dd, $^3J_{4,5}$ = 2.6 Hz, $^3J_{3,4}$ = 6.1 Hz, 1 H, 4-H), 5.56 (dd, $^3J_{4,5}$ = 2.6 Hz, $^3J_{5,6}$ = 8.6 Hz, 1 H, 5-H), 5.79 (d, $^3J_{4',5'}$ =

7.1 Hz, 1 H, 5'-H), 5.99 (dd, $^3J_{2,3}$ = 1.5 Hz, $^2J_{2\text{-F},2}$ = 48.3 Hz, 1 H, 2-H), 7.26-7.46 (m, 5 H, CH-Ar), ^{13}C-NMR (100 MHz, CDCl$_3$, 25 °C): δ = 14.4 (4'-CH$_3$), 20.8, 20.9, 21.1 (4 OAc), 55.6 (4'-C), 61.9 (7-C), 68.2 (6-C), 68.5 (5-C), 70.6 (d, $^2J_{2\text{-F},3}$ = 19.2 Hz, 3-C), 70.9 (d, $^3J_{2\text{-F},4}$ = 3.2 Hz, 4-C), 80.6 (5'-C), 90.2 (d, $^1J_{2\text{-F},2}$ = 183.1 Hz, 2-C), 125.7, 129.0, 129.2 (CH-Ar), 132.6 (Cq-Ar), 153.0 (2'-C), 165.8 (d, $^2J_{2\text{-F},1}$ = 24.0 Hz, 1-C), 170.0, 170.7, 170.9, 171.3 (4 CO-OAc), ^{19}F NMR (CDCl$_3$, 565 MHz, 25°C): δ = -209.99; HRMS (ESI): calcd. for C$_{25}$H$_{30}$FNNaO$_{12}$ [M + Na]$^+$ 578.1650, found 578.1640.

4 NMR Spectra[1,2]

2-azido-2-deoxy-D-glycero-D-ido-heptose (6a)

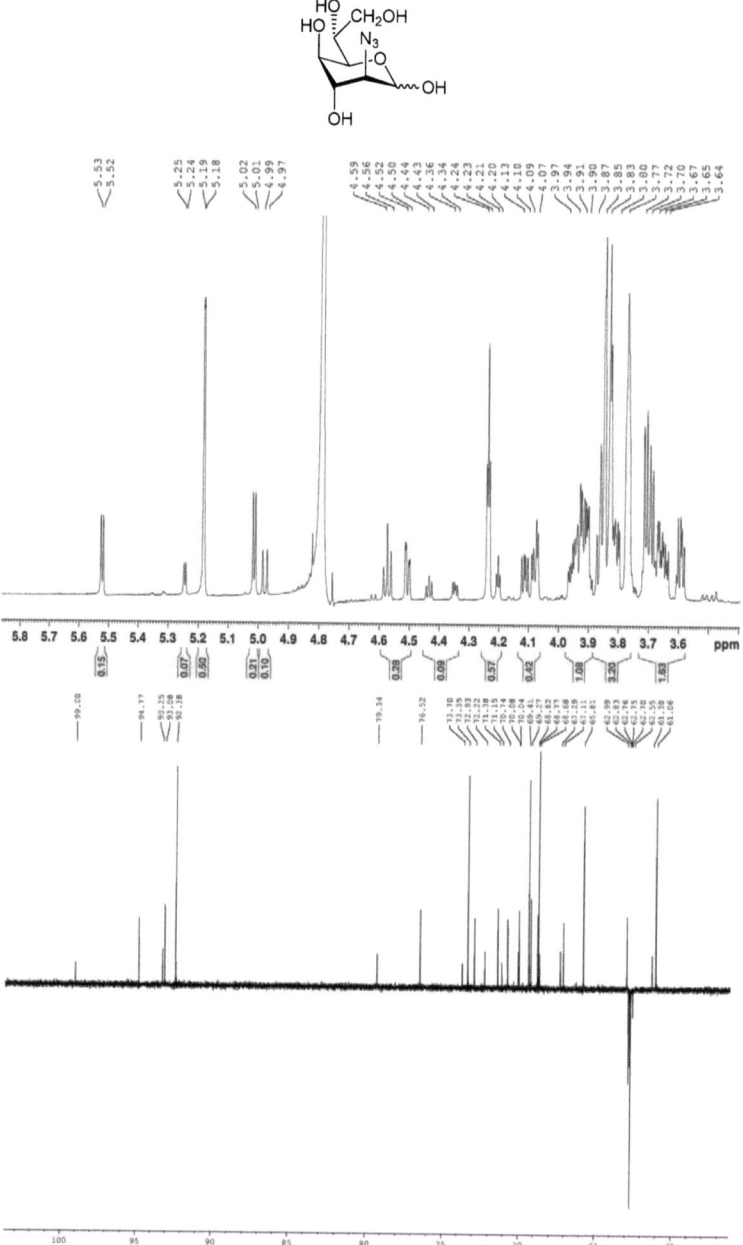

2-azido-2-deoxy-D-threo-L-galacto-octose (6b)

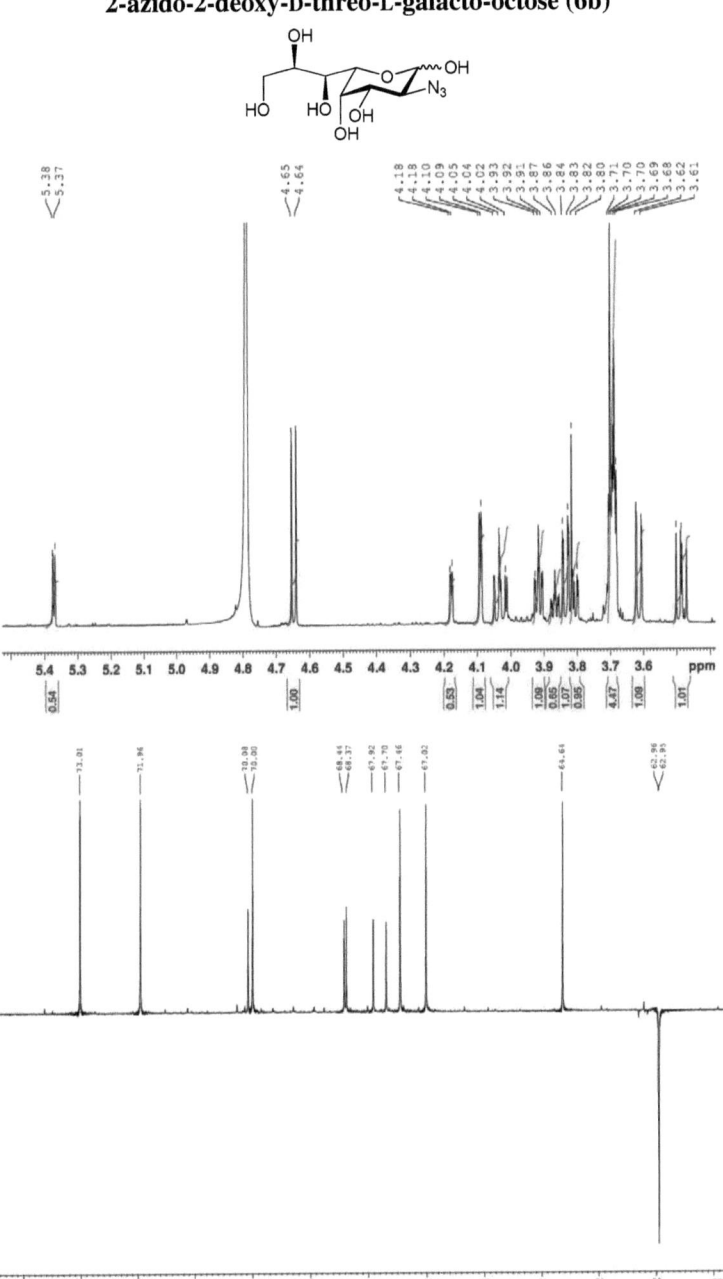

2-azido-2-deoxy-D-erythro-L-galacto-octose (6c)

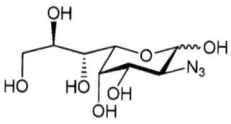

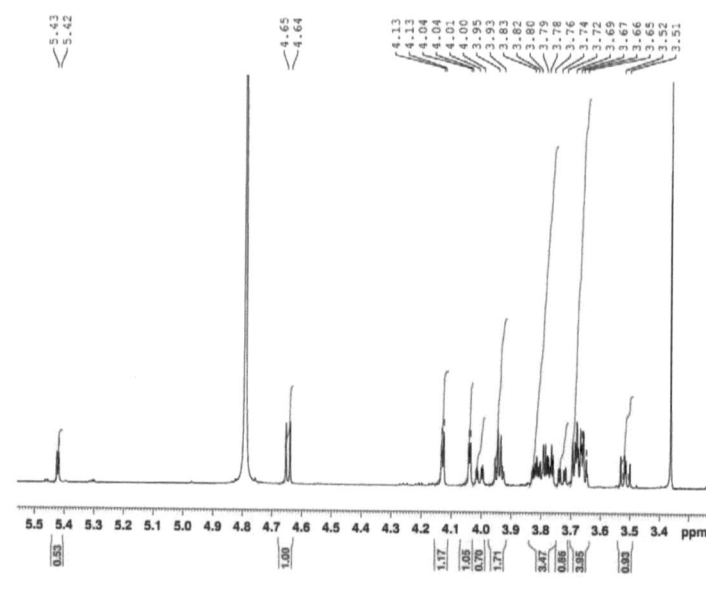

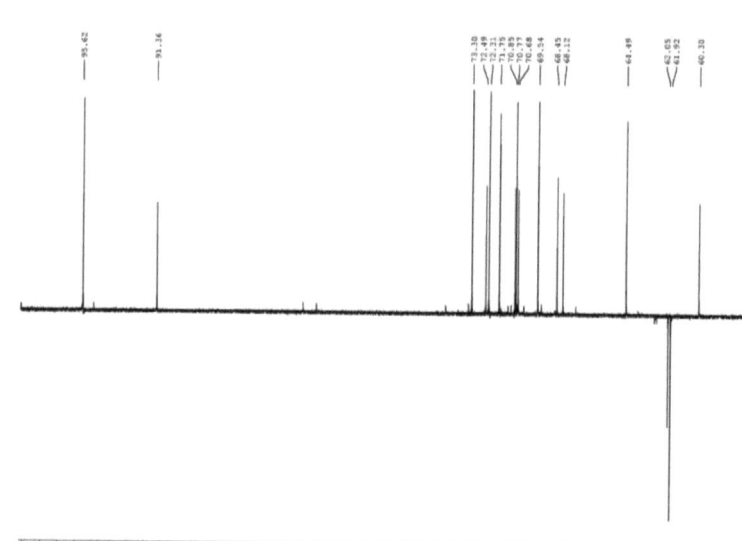

2-acetamido-1,3,4,6,7-penta-O-acetyl-2-deoxy-D-glycero-D-ido-heptose (7a)

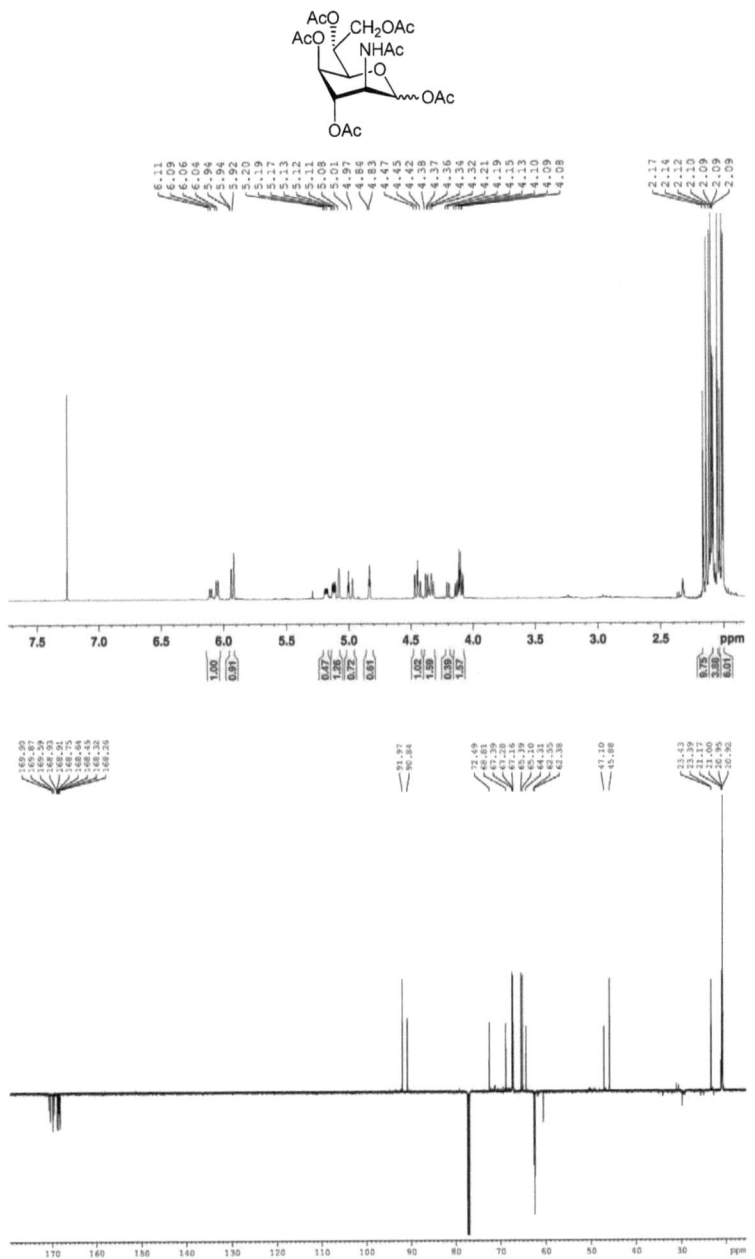

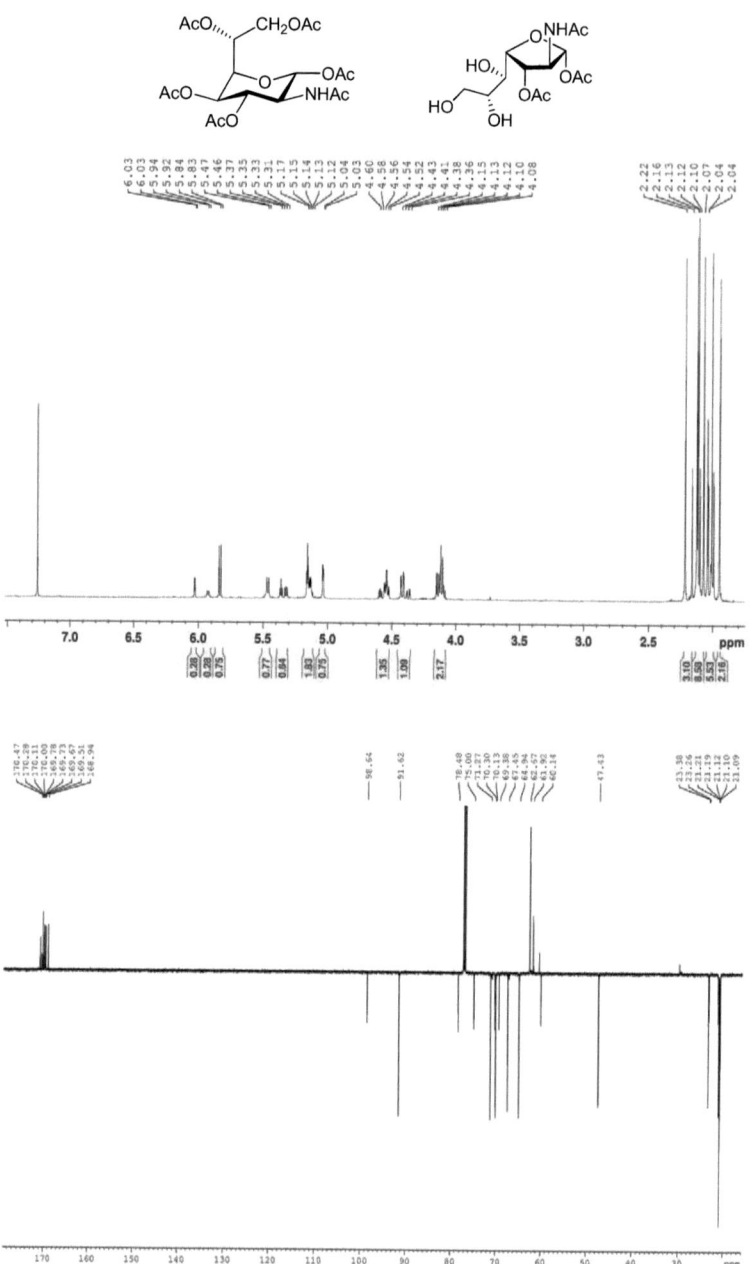

2-acetamido-1,3,4,6,7,8-hexa-O-acetyl-2-deoxy-D-threo-L-galacto-octose (7b)

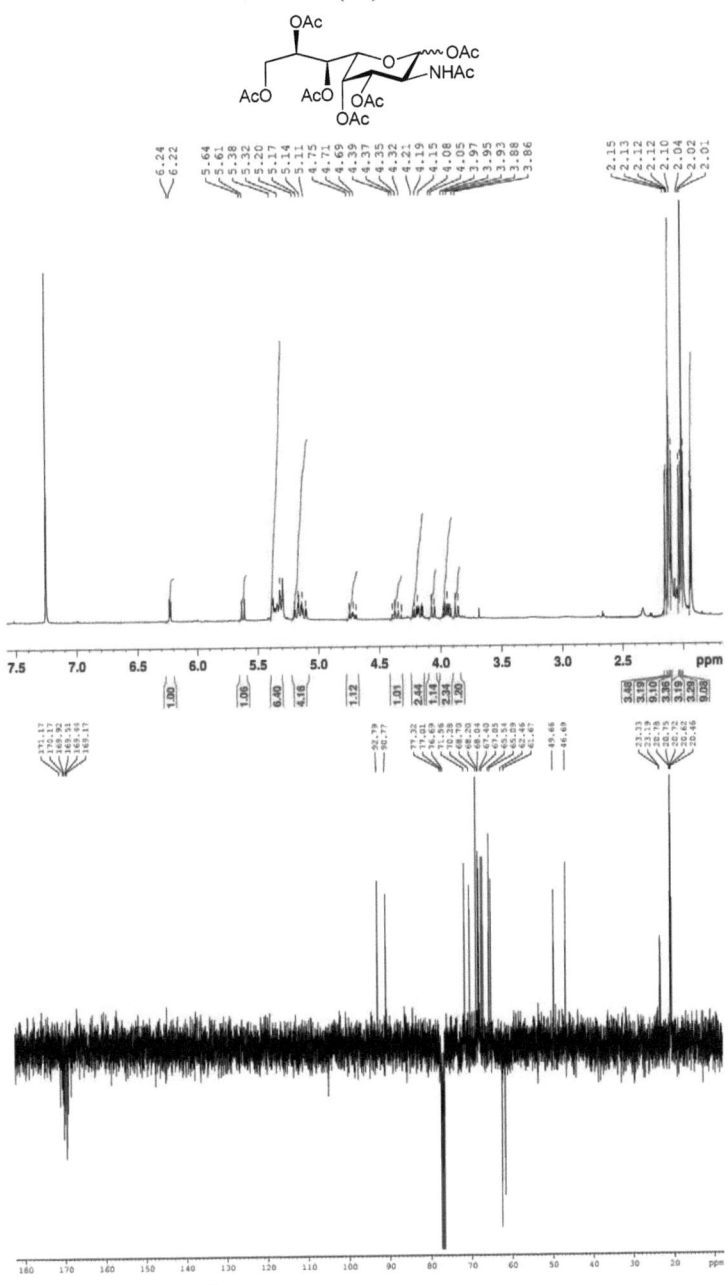

2-acetamido-1,3,4,6,7,8-hexa-O-acetyl-2-deoxy-D-erythro-L-galacto-octose (7c)

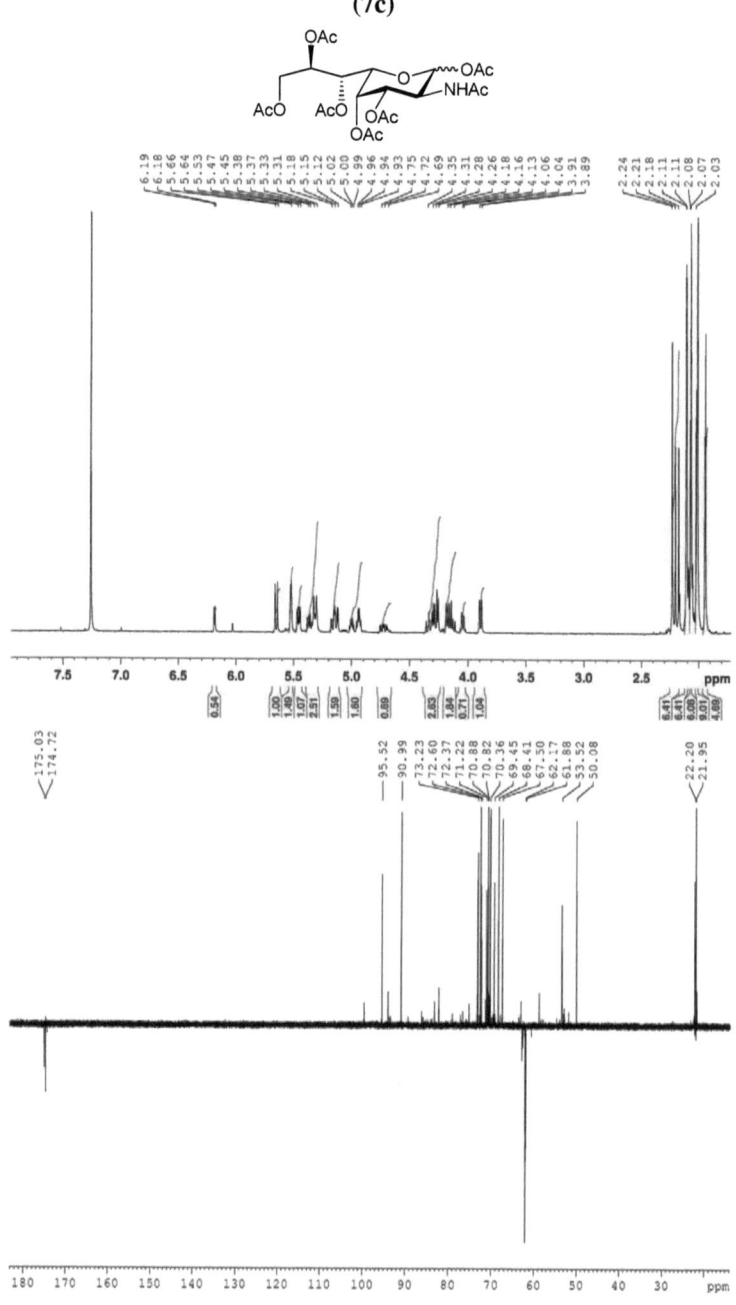

2-acetamido-2-deoxy-D-glycero-D-ido-heptose (8a)

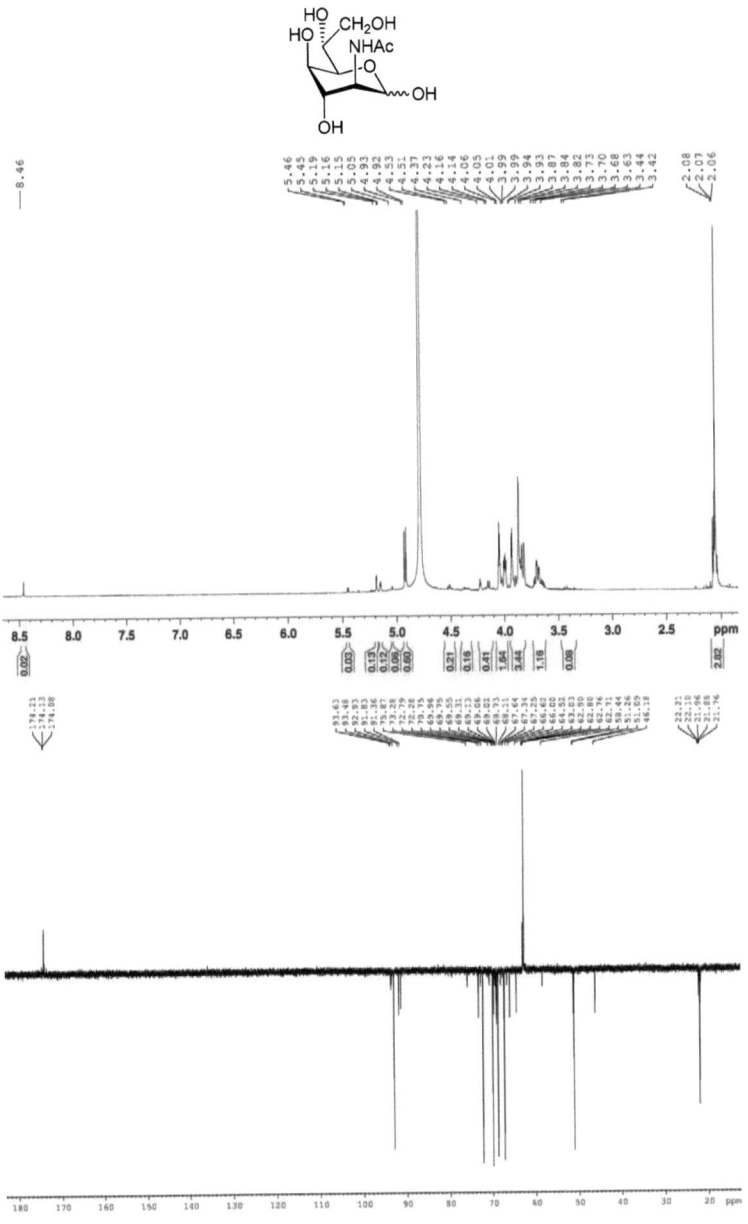

2-acetamido-2-deoxy-D-threo-L-galacto-octose (8b)

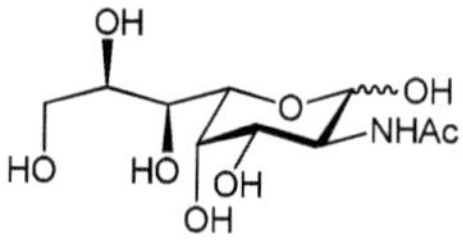

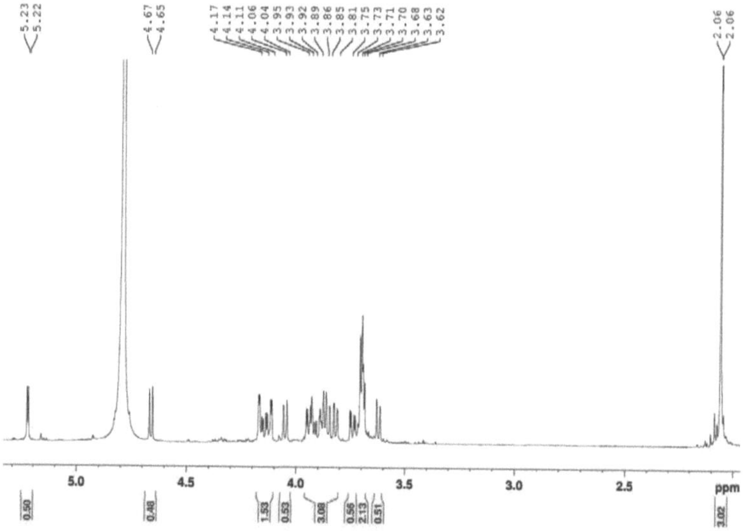

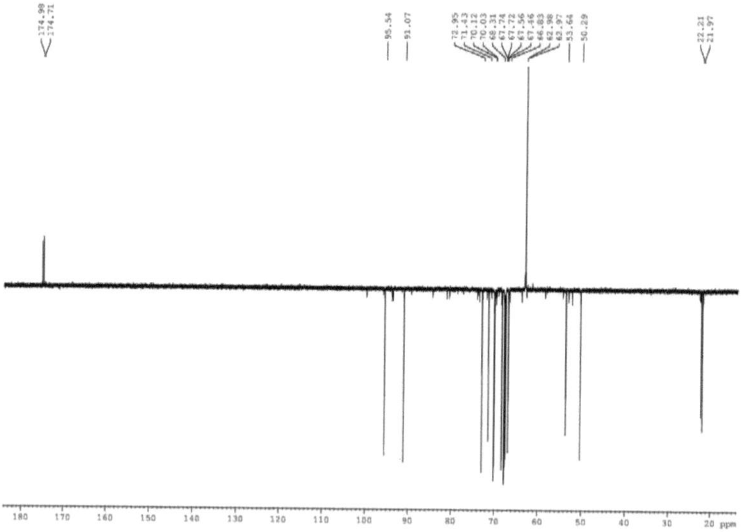

2-acetamido-2-deoxy-D-erythro-L-galacto-octose (8c)

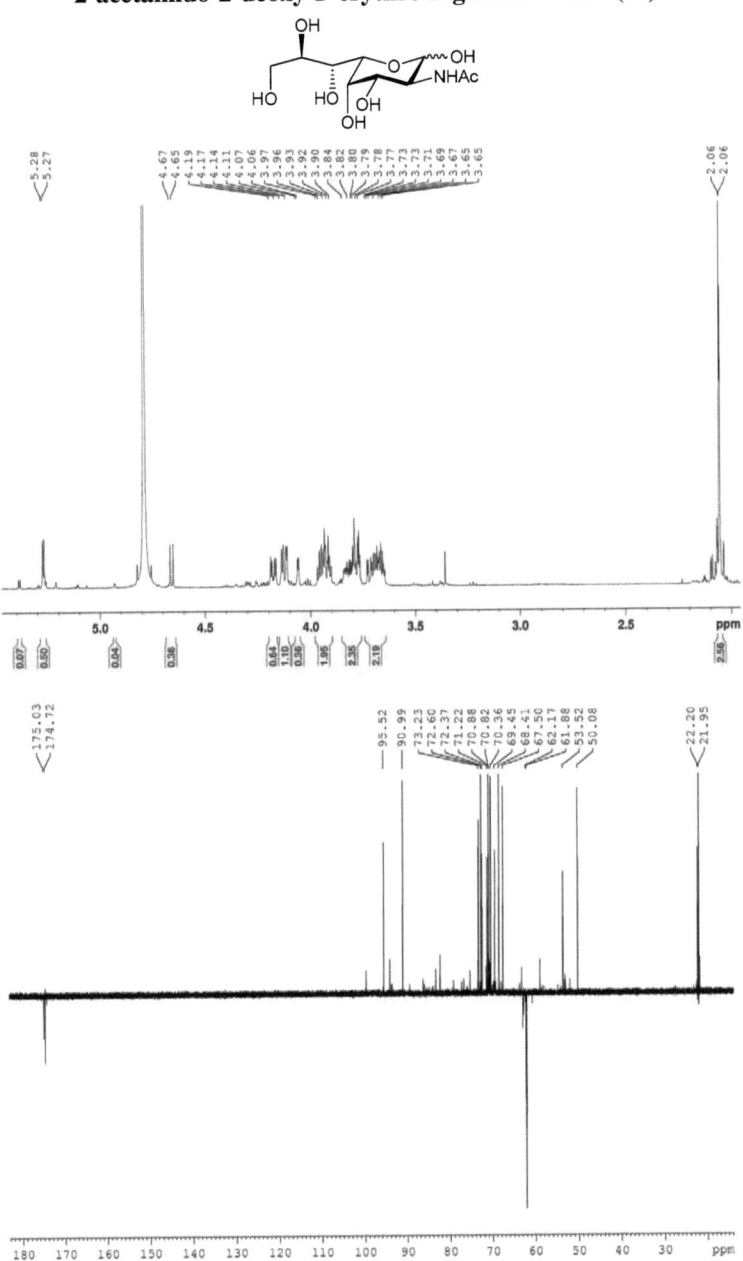

(R,E)-methyl 4-(dibenzylamino)-5-((4-methoxybenzyl)oxy)pent-2-enoate (9a)

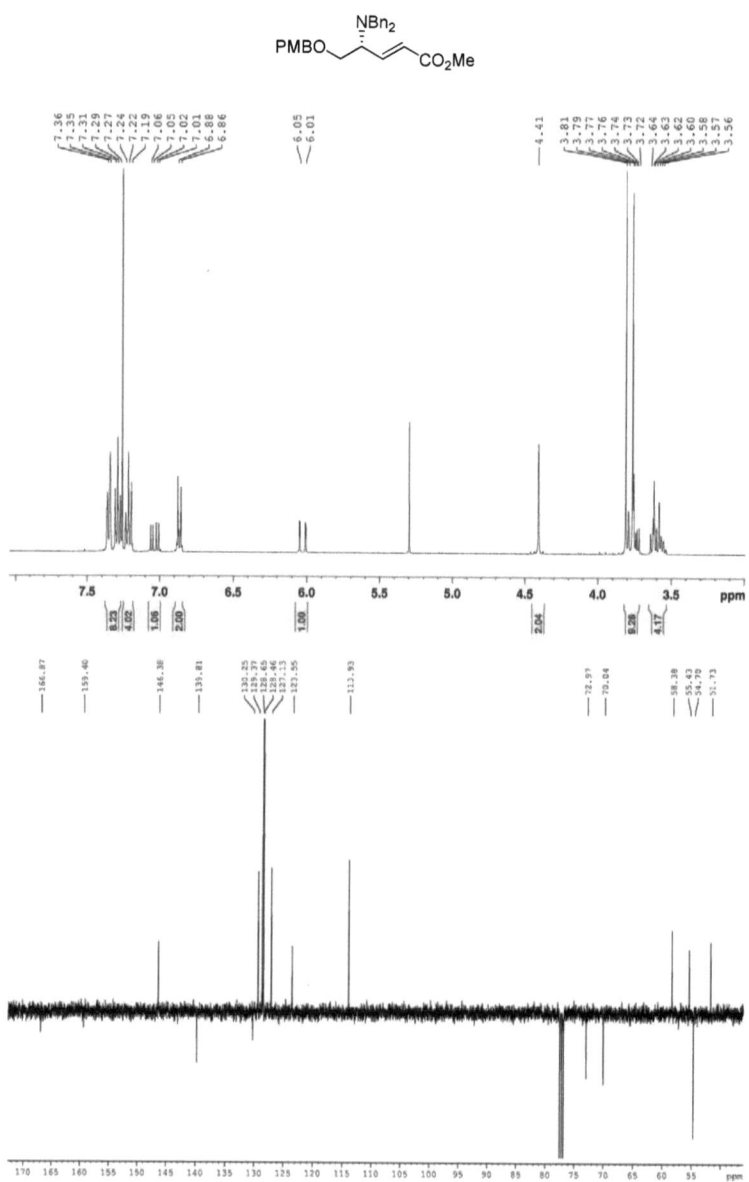

(R,E)-methyl 4-(dibenzylamino)-5-(pivaloyloxy)pent-2-enoate (9c)

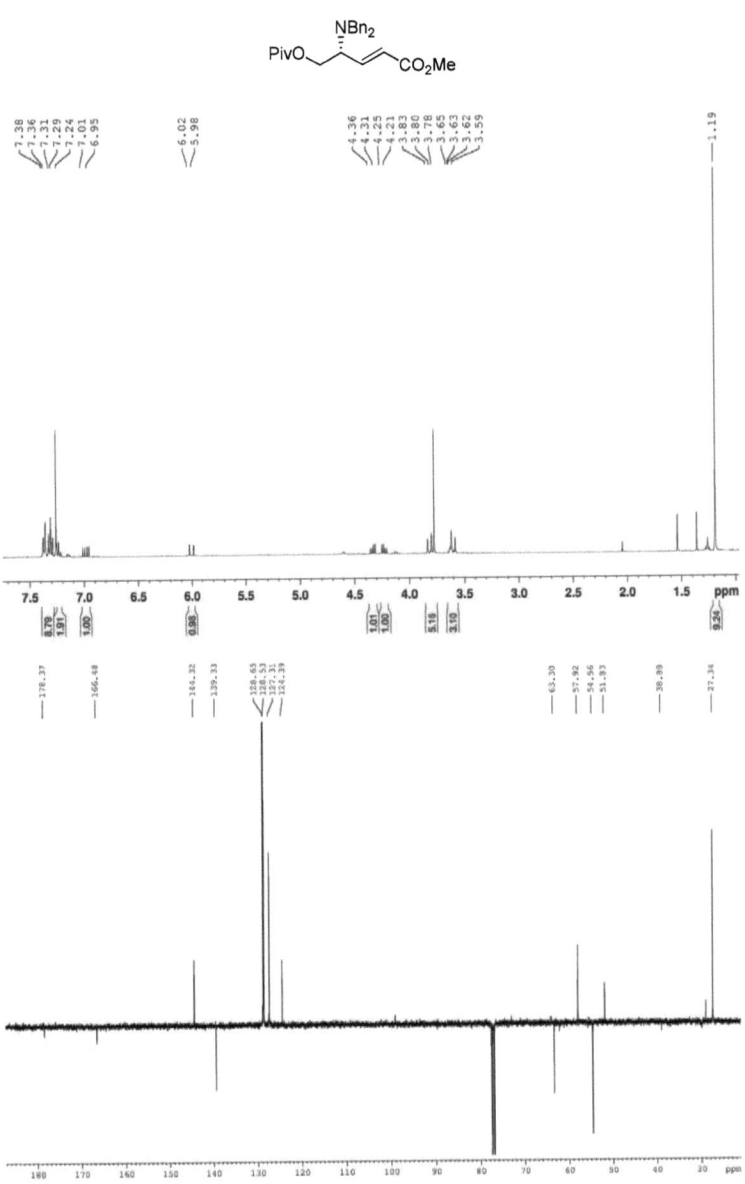

((2S,3S)-3-((S)-1-(dibenzylamino)-2-((4-methoxybenzyl)oxy)ethyl)oxiran-2-yl)methanol (10)

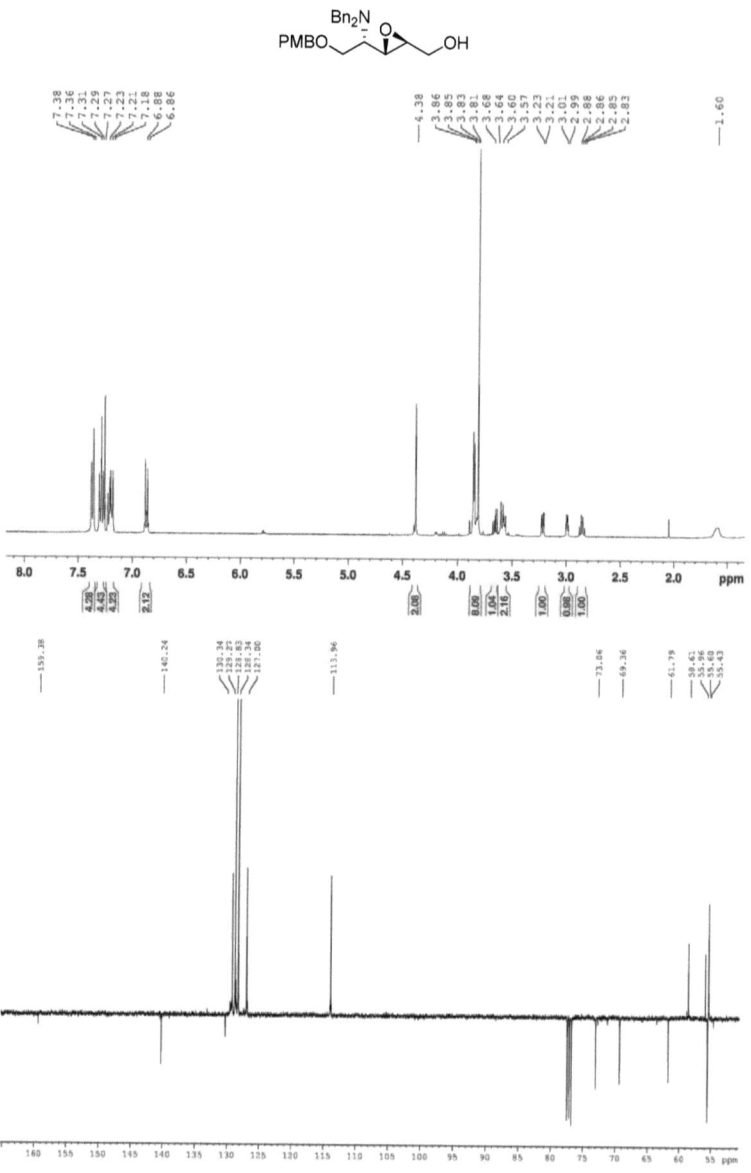

(2R,3S)-methyl 3-((S)-2-((tert-butyldiphenylsilyl)oxy)-1-(dibenzylamino)ethyl)oxirane-2-carboxylate (11)

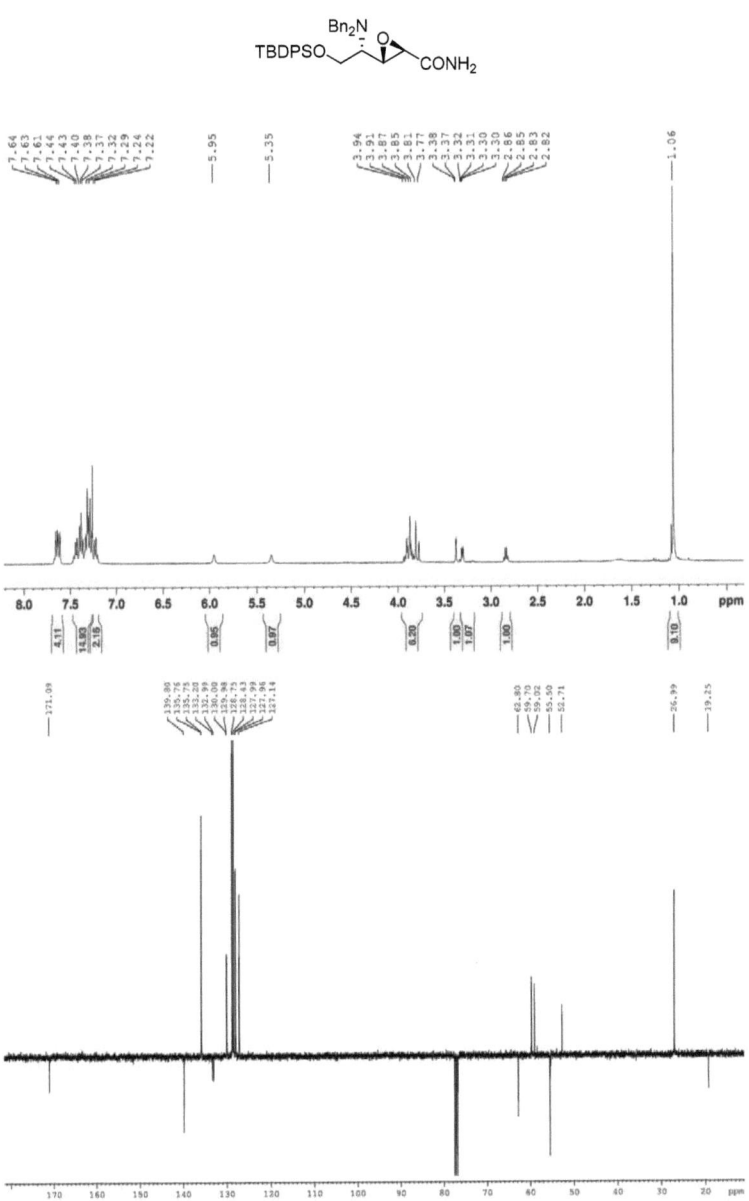

(R)-tert-butyl 4-((1R,2S,E)-1-(benzyloxy)-5-ethoxy-2-fluoro-5-oxopent-3-en-1-yl)-2,2-dimethyloxazolidine-3-carboxylate (13)

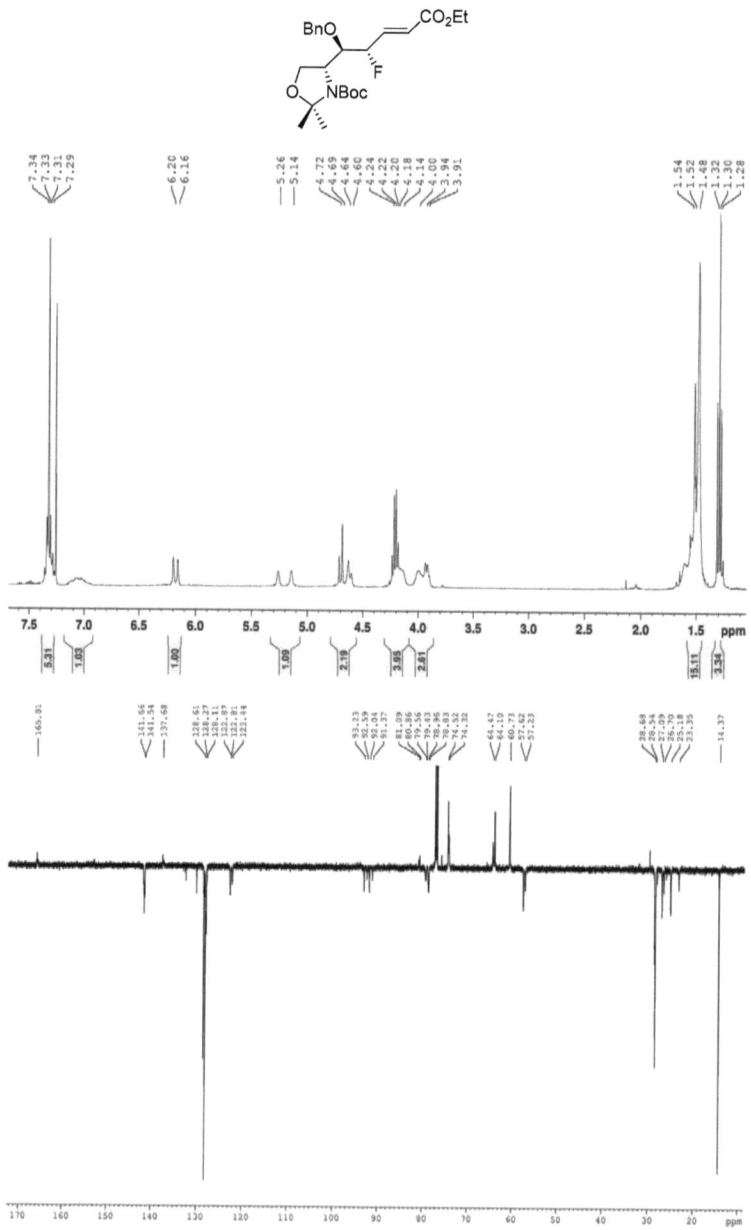

(4S,5R)-3-(2-fluoroacetyl)-4-methyl-5-phenyloxazolidin-2-one (14)

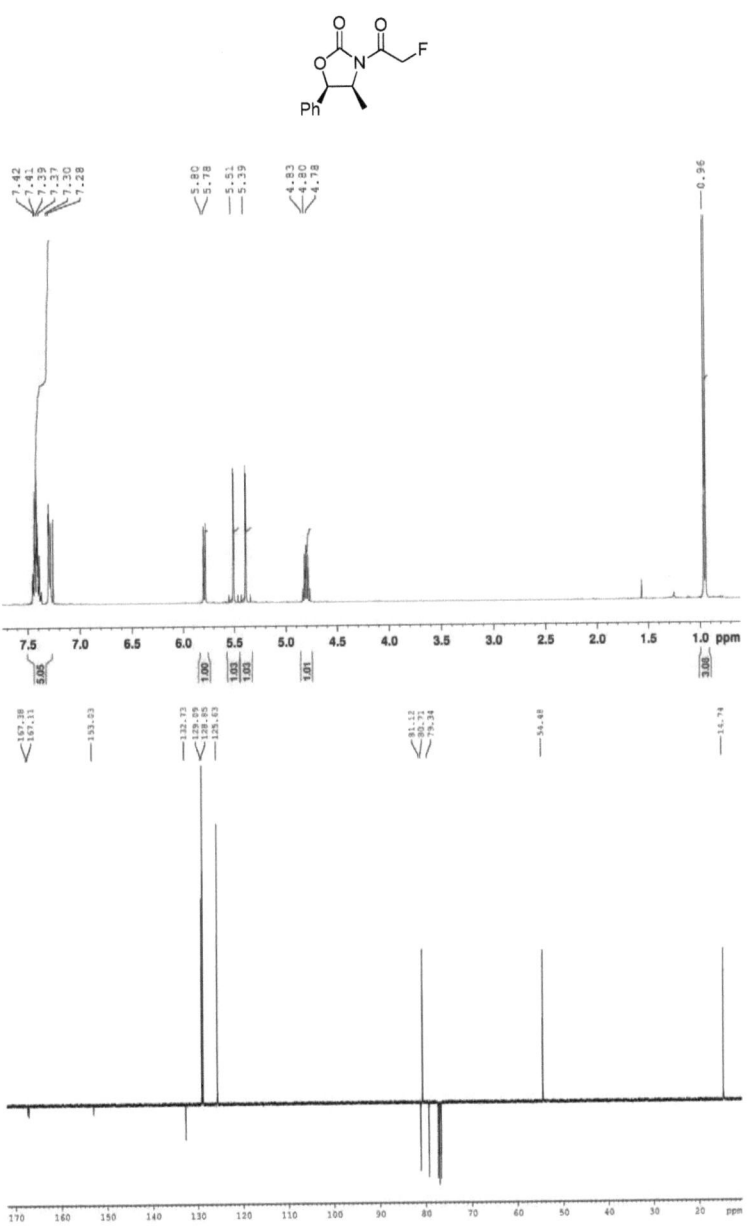

(S)-tert-butyl-4-((1S,2R)-2-fluoro-1-hydroxy-3-((4S,5R)-4-methyl-2-oxo-5-phenyloxazolidin-3-yl)-3-oxopropyl)-2,2-dimethyloxazolidine-3-carboxylate (15)

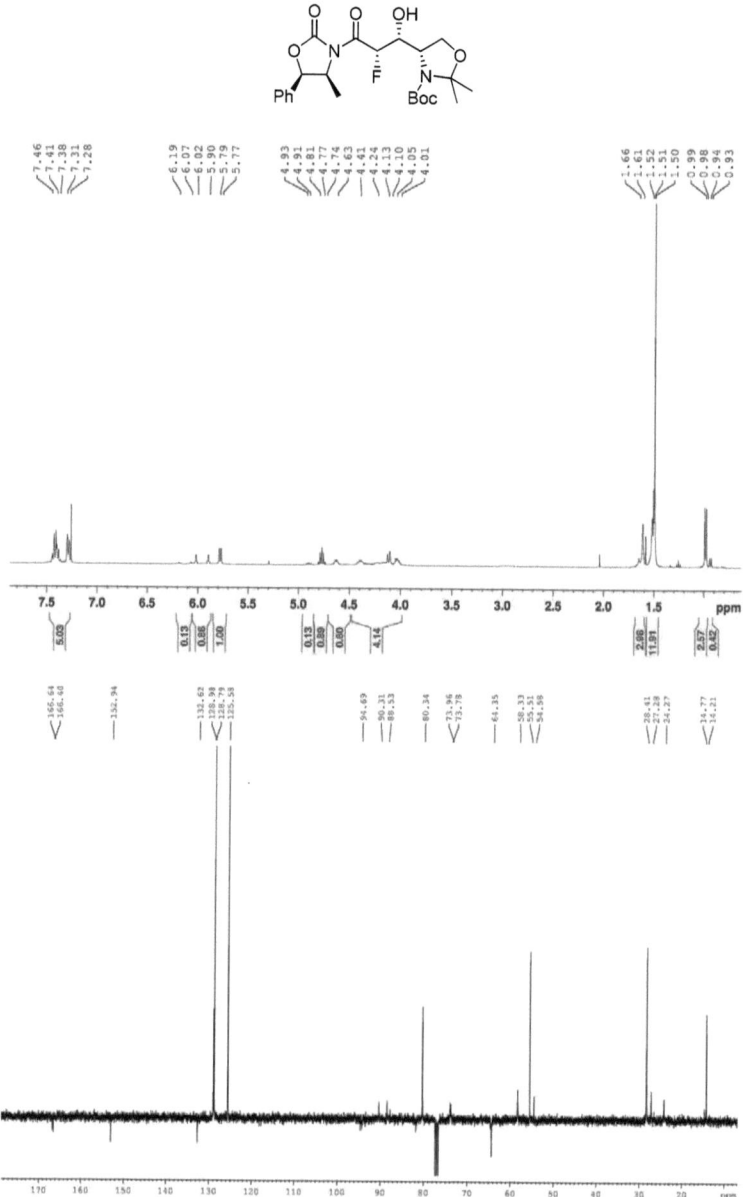

(S)-tert-butyl-4-((1R,2S)-2-fluoro-1-hydroxy-3-((4S,5R)-4-methyl-2-oxo-5-phenyloxazolidin-3-yl)-3-oxopropyl)-2,2-dimethyloxazolidine-3-carboxylate (16)

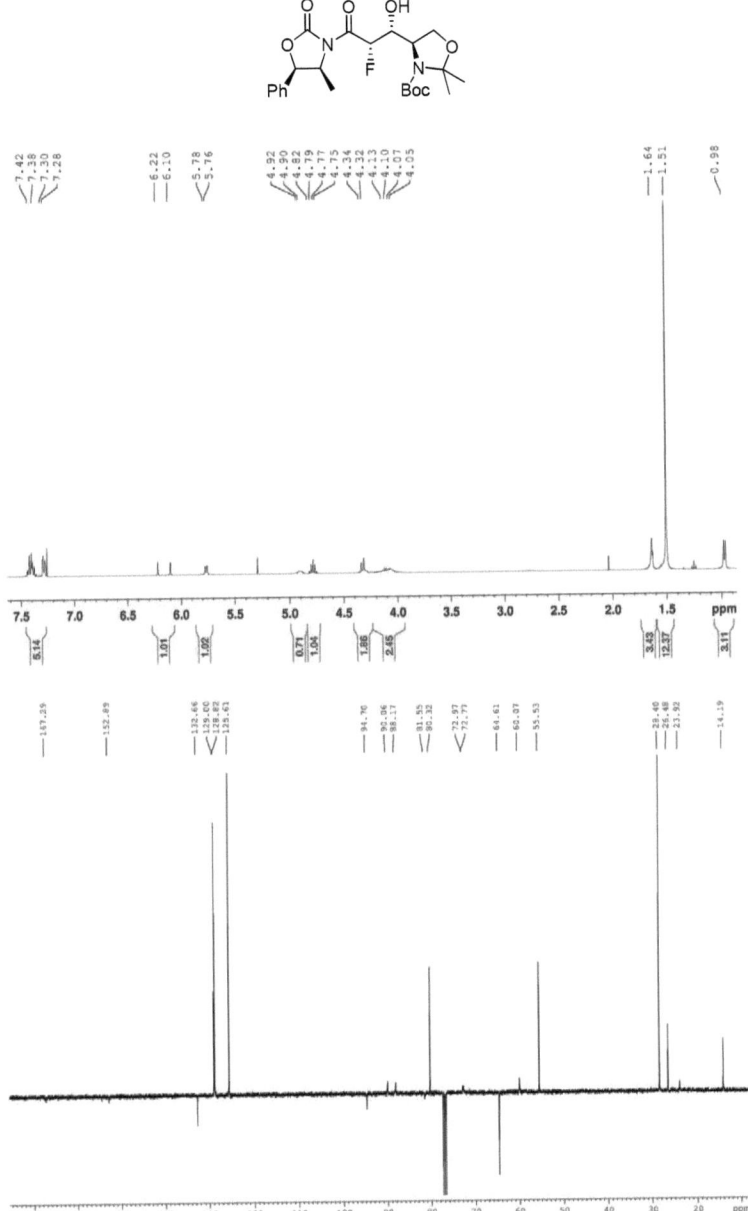

4R,5R)-tert-butyl-4-((1R,2S)-2-fluoro-1-hydroxy-3-((4S,5R)-4-methyl-2-oxo-5-phenyloxazolidin-3-yl)-3-oxopropyl)-2,2,5-trimethyloxazolidine-3-carboxylate (17)

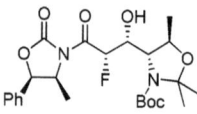

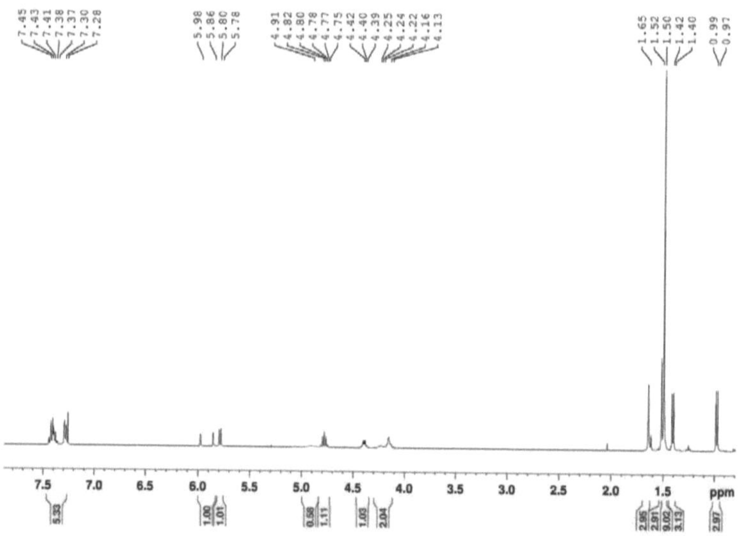

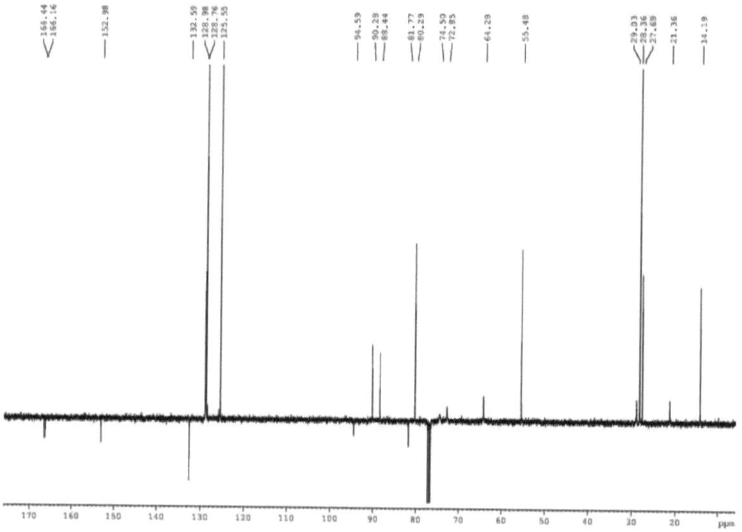

(4S,5S)-tert-butyl 4-((1R,2S)-2-fluoro-1-hydroxy-3-((4S,5R)-4-methyl-2-oxo-5-phenyloxazolidin-3-yl)-3-oxopropyl)-2,2,5-trimethyloxazolidine-3-carboxylate (18)

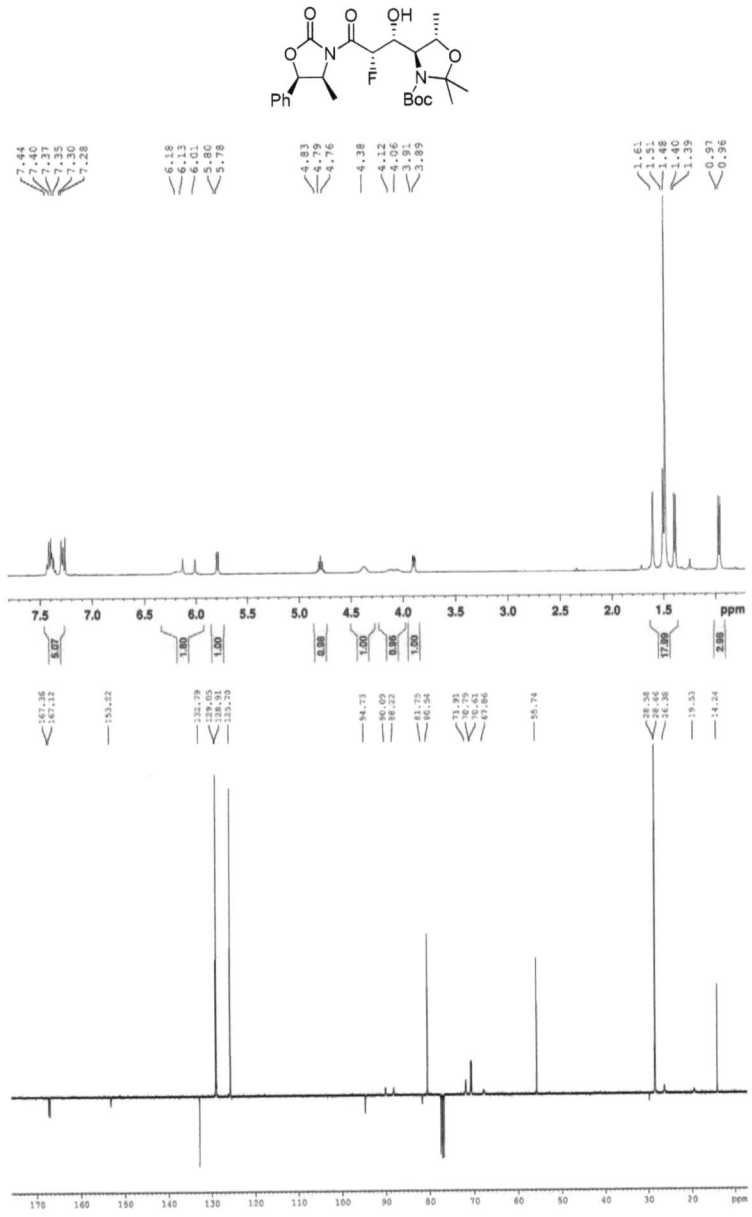

(4S,5R)-tert-butyl-4-((1R,2S)-2-fluoro-1-hydroxy-3-((4S,5R)-4-methyl-2-oxo-5-phenyloxazolidin-3-yl)-3-oxopropyl)-2,2,5-trimethyloxazolidine-3-carboxylate (19)

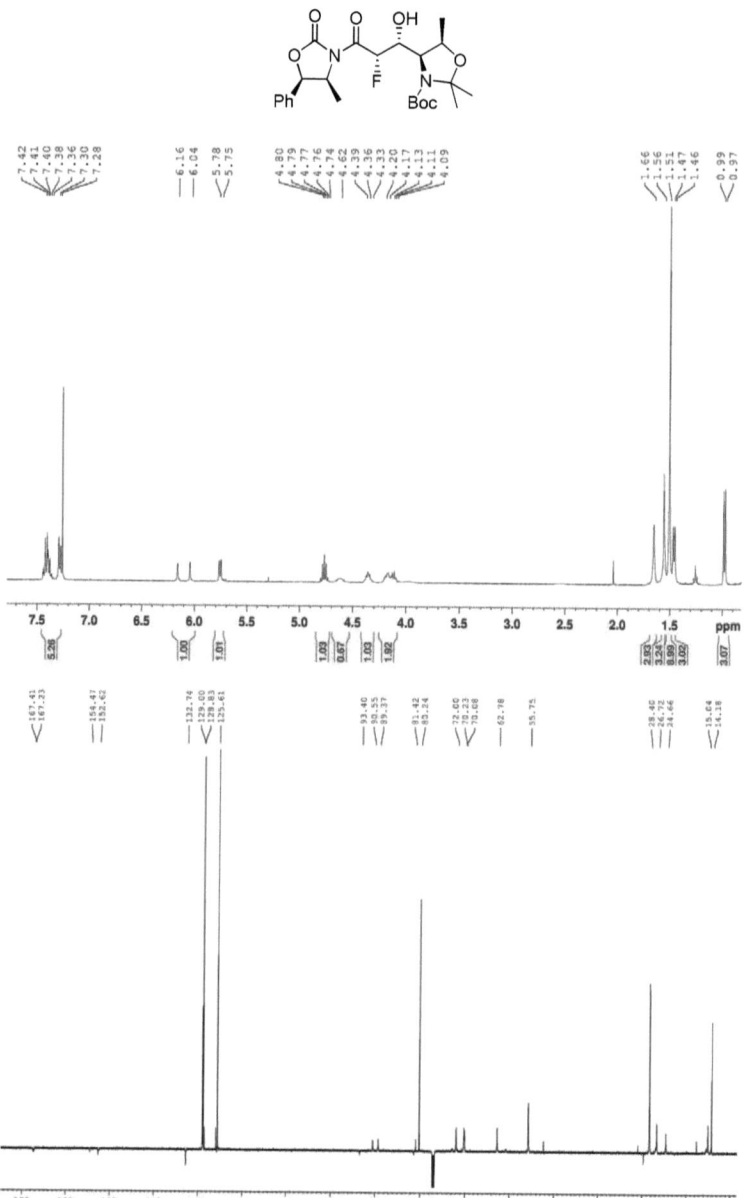

(2S,3R,4S)-methyl-4-((tert-butoxycarbonyl)amino)-2-fluoro-3,5-dihydroxypentanoate (20)

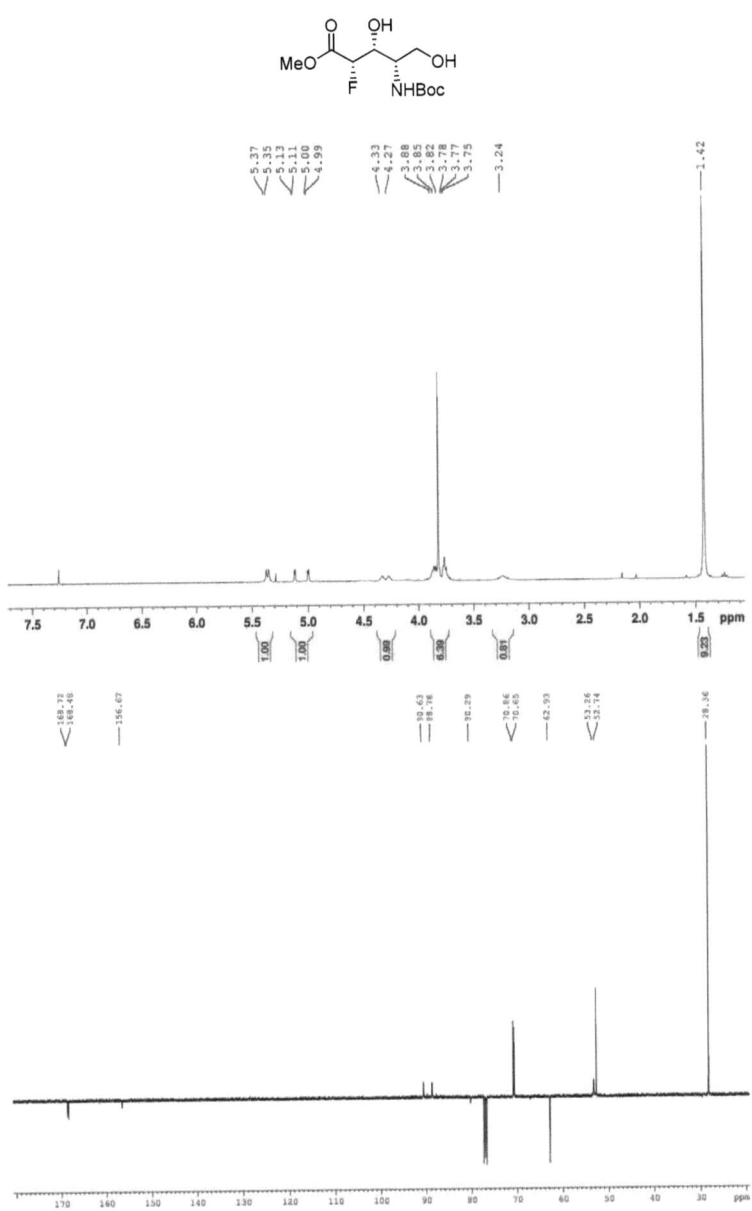

(2S,3R,4R)-methyl-4-((tert-butoxycarbonyl)amino)-2-fluoro-3,5-dihydroxypentanoate (21)

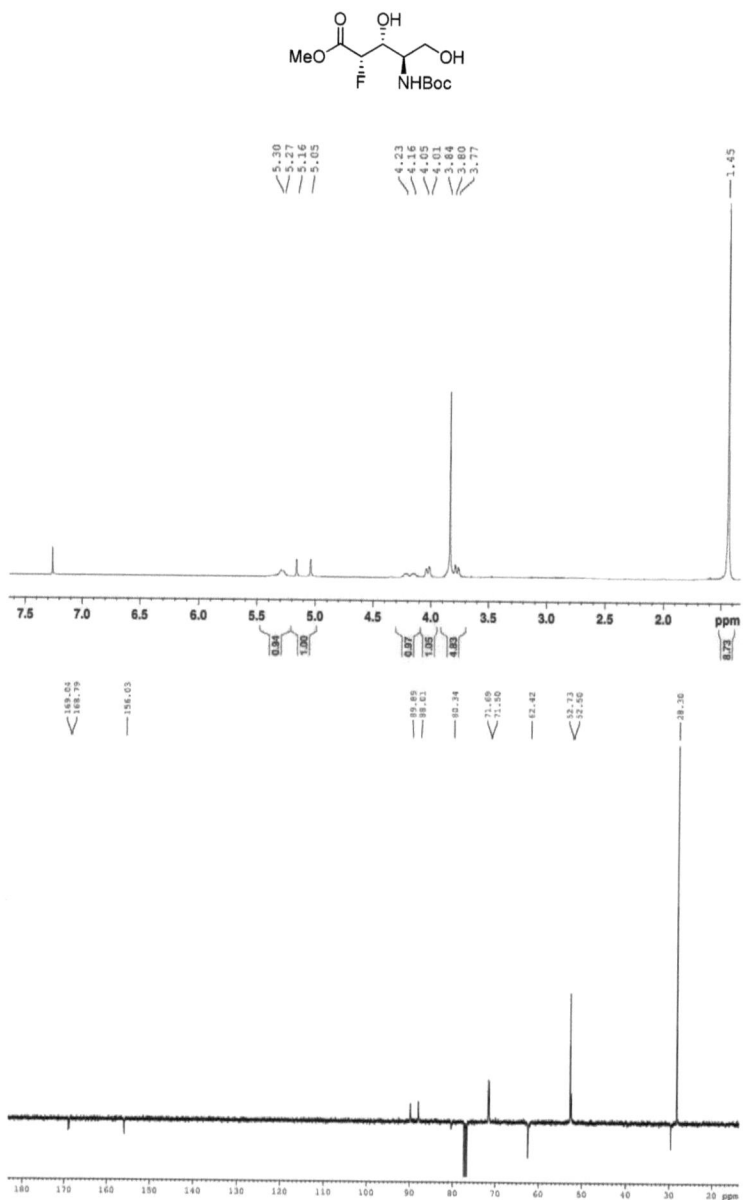

(2S,3R,4S,5R)-methyl-4-((tert-butoxycarbonyl)amino)-2-fluoro-3,5-dihydroxyhexanoate (22)

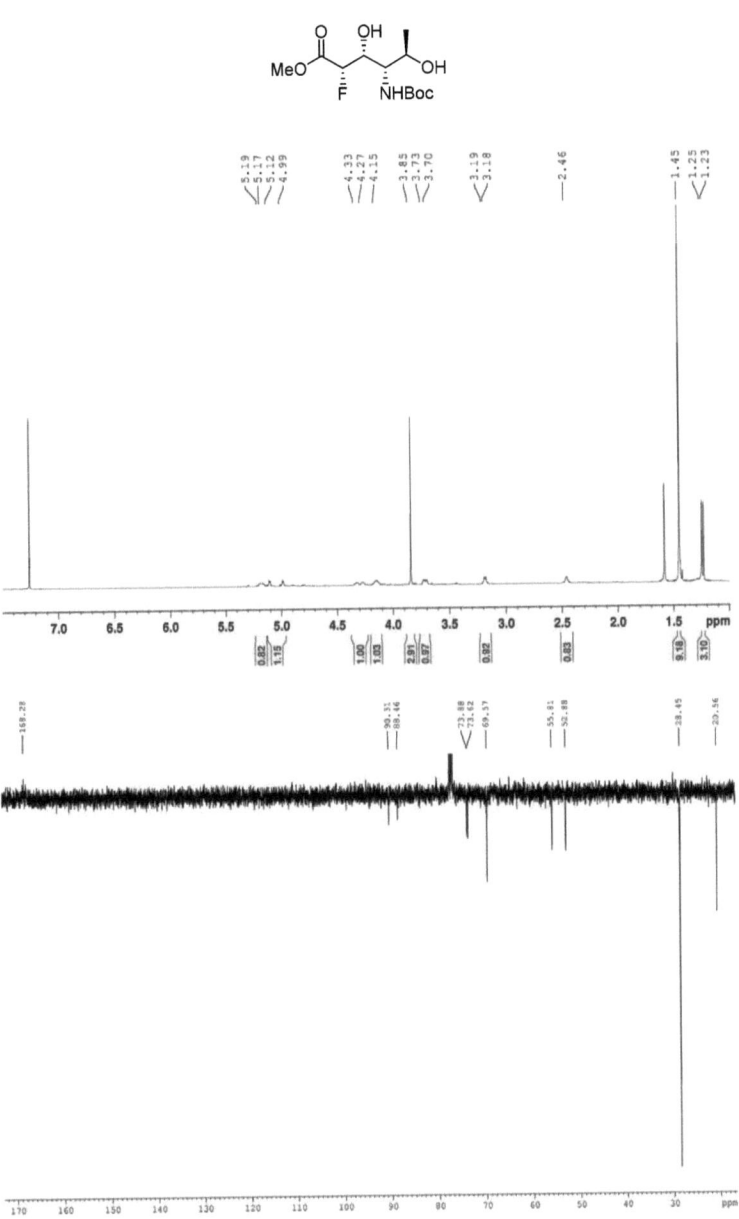

(2S,3R,4R,5S)-methyl-4-((tert-butoxycarbonyl)amino)-2-fluoro-3,5-dihydroxyhexanoate (23)

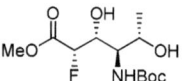

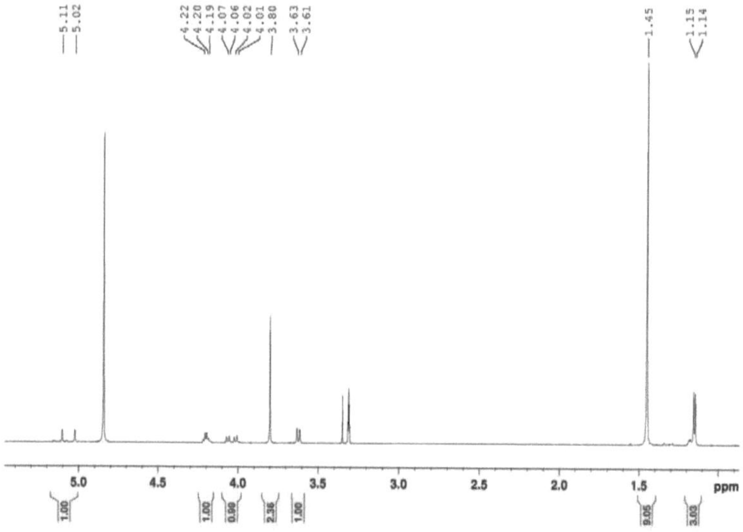

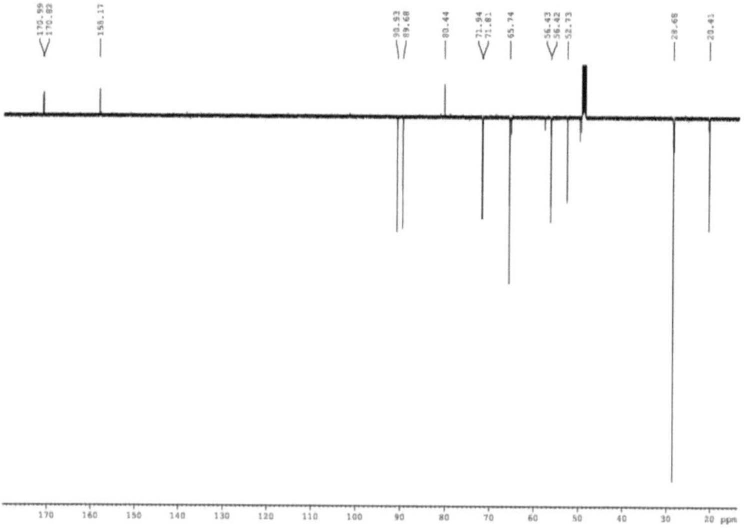

4-acetamido-1,3-di-O-acetyl-2,4-dideoxy-2-fluoro-D-xylose (24)

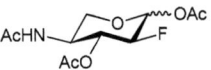

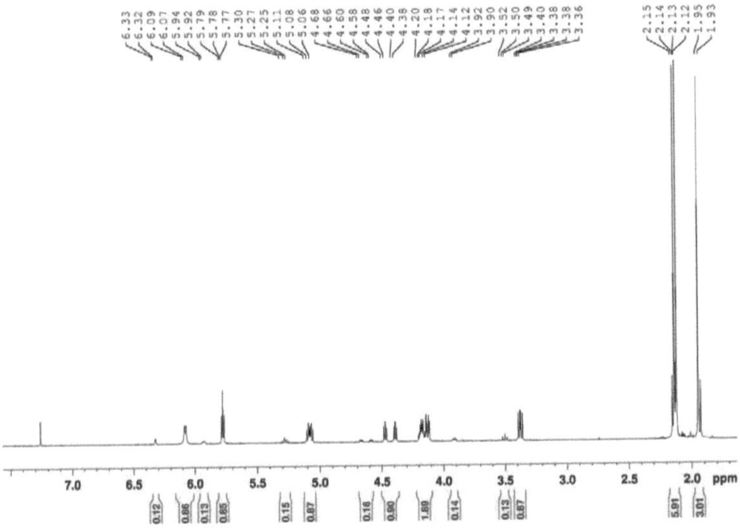

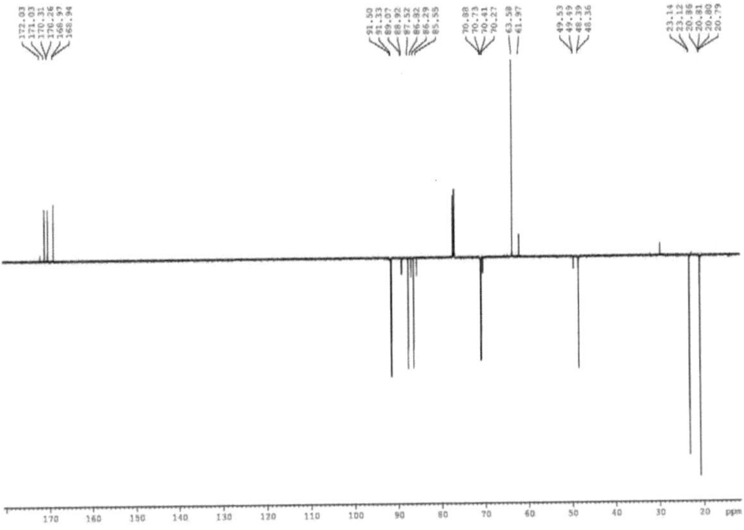

4-acetamido-1-O-acetyl-2,4-dideoxy-2-fluoro-D-arabinose (25)

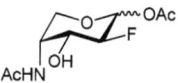

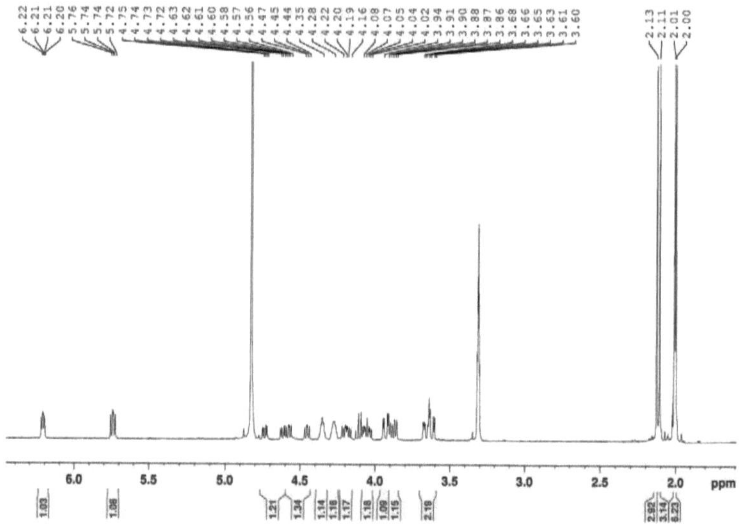

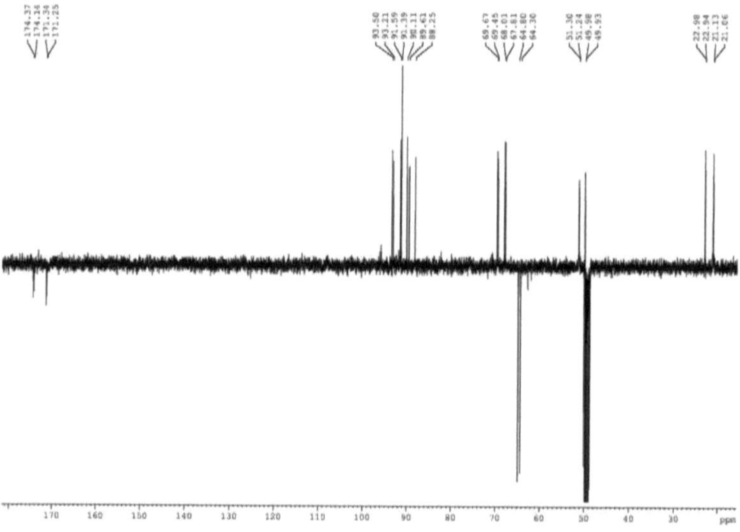

4-acetamido-1,3-di-O-acetyl-2,4,6-trideoxy-2-fluoro-D-idose (26)

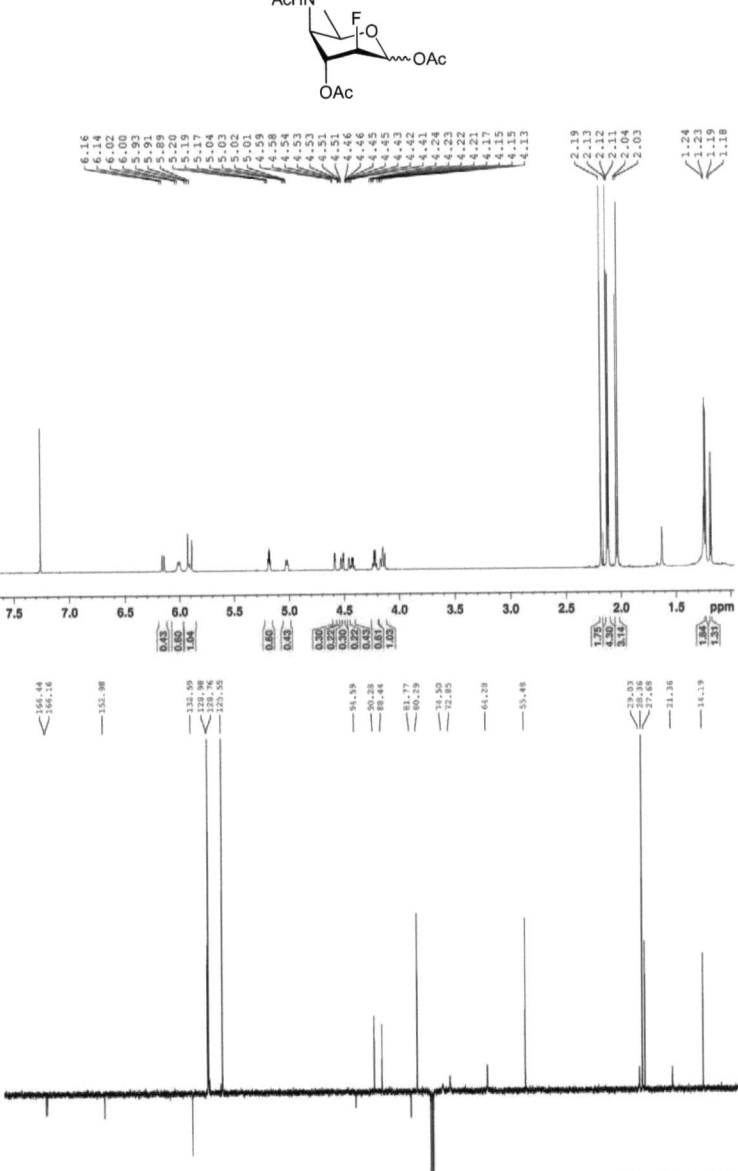

4-acetamido-1-O-acetyl-2,4,6-trideoxy-2-fluoro-L-galactose (27)

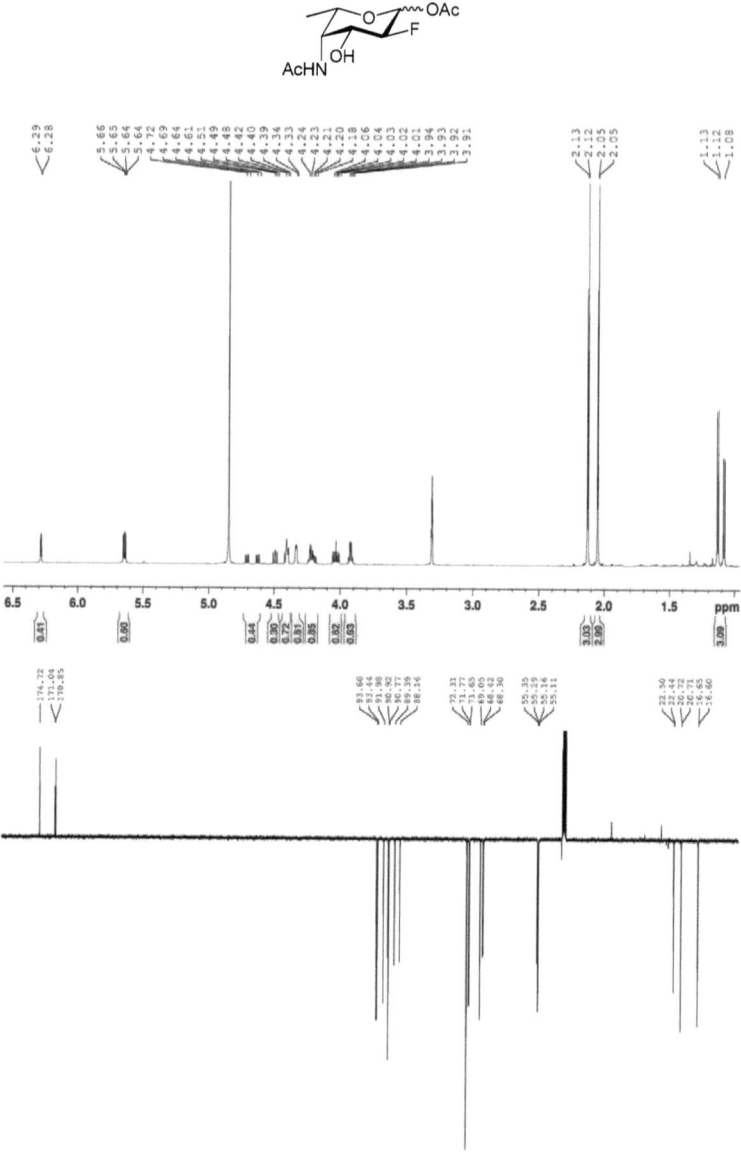

4-acetamido-2,4-dideoxy-2-fluoro-D-xylose (28)

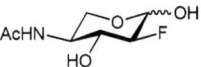

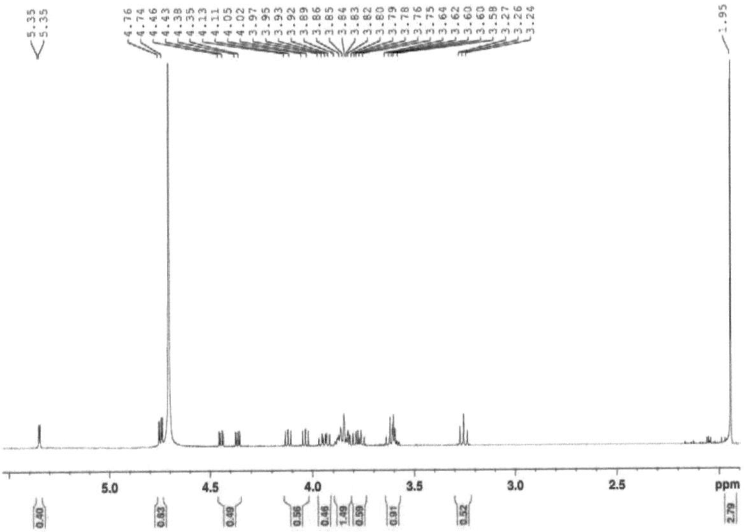

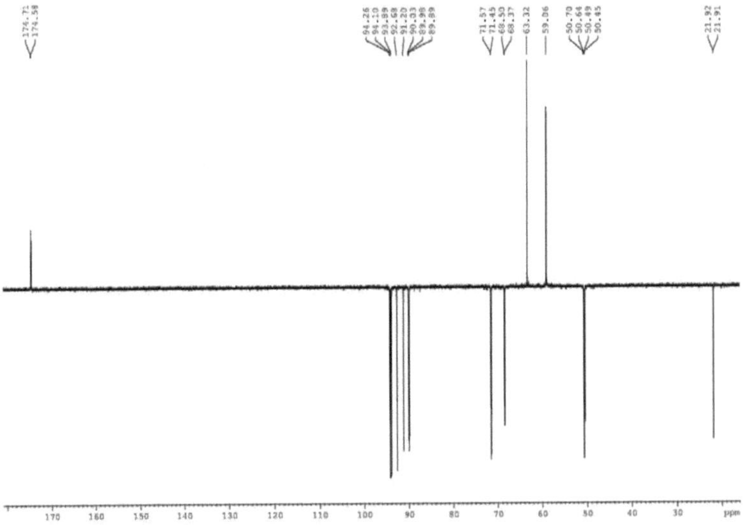

4-acetamido-2,4-dideoxy-2-fluoro-D-arabinose (29)

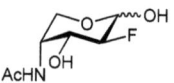

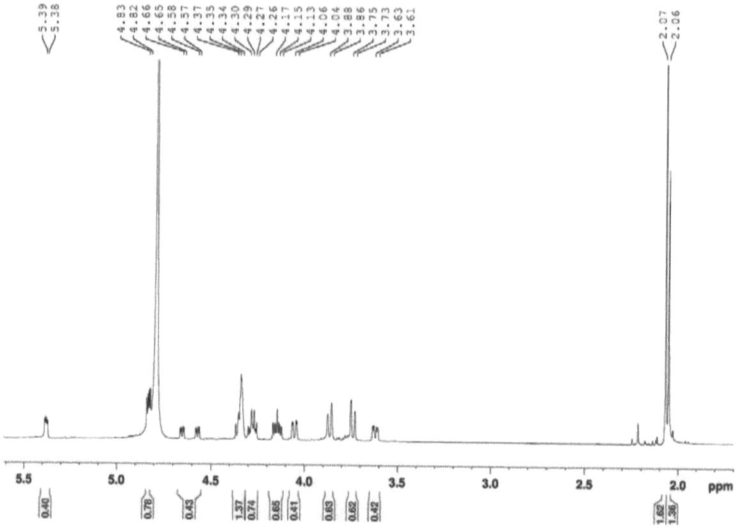

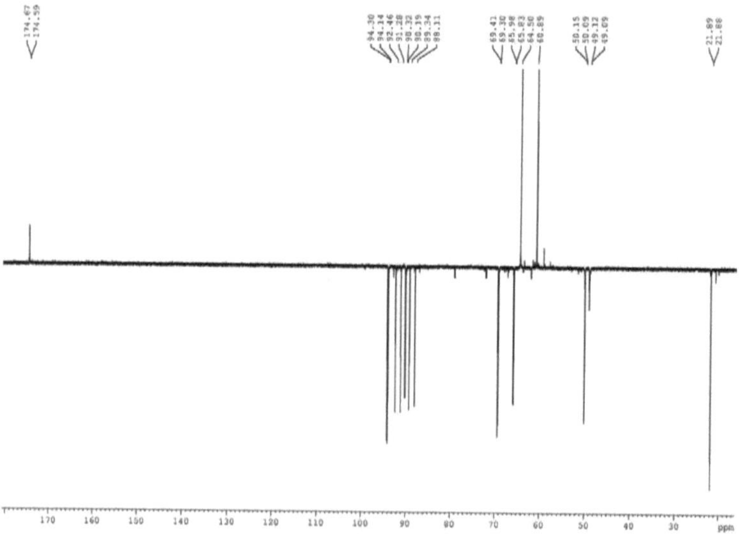

4-acetamido-2,4,6-trideoxy-2-fluoro-D-idose (30)

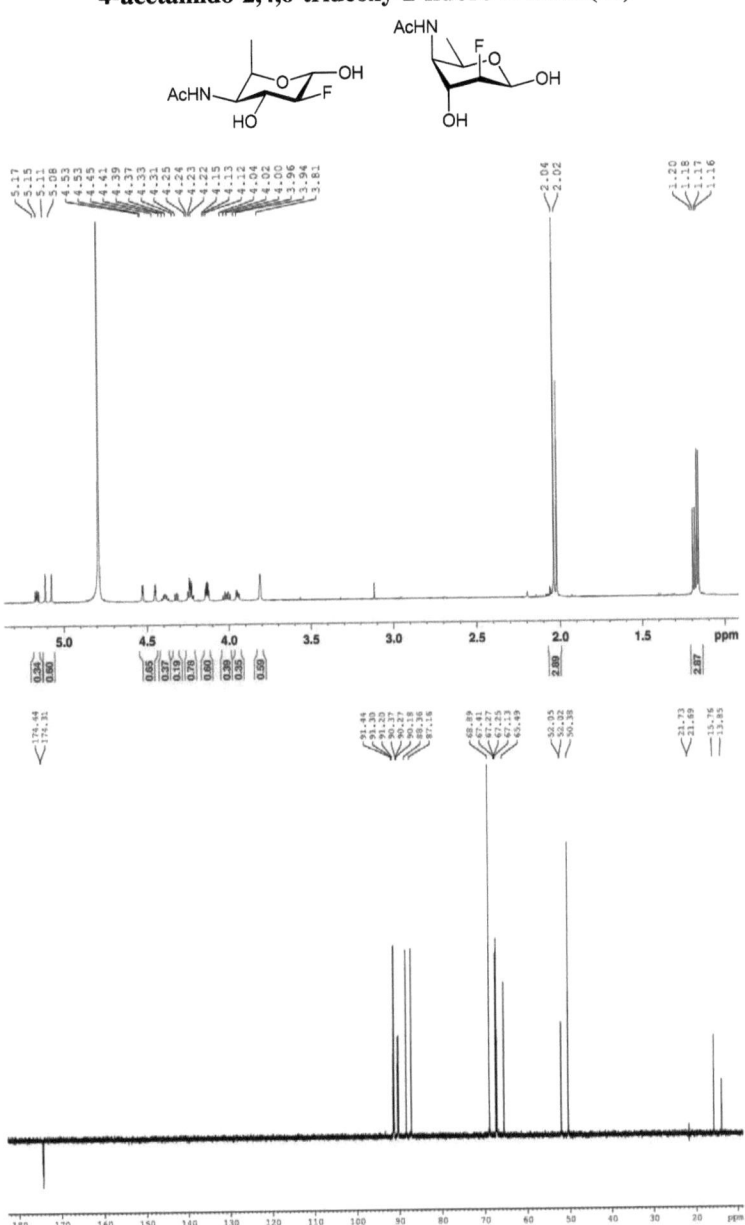

4-acetamido-2,4,6-trideoxy-2-fluoro-D-galactose (31)

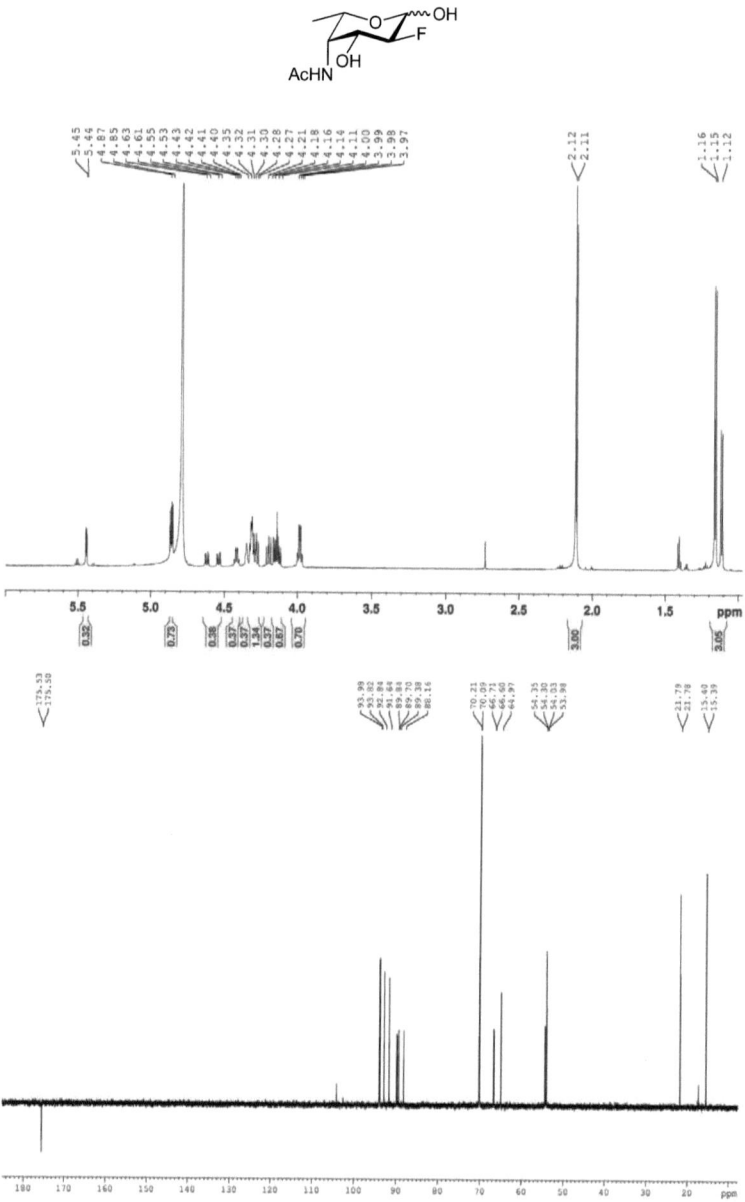

(S)-tert-butyl 4-((1R,2S)-2-fluoro-1-hydroxy-3-((4S,5R)-4-methyl-2-oxo-5-phenyloxazolidin-3-yl)-3-oxopropyl)-2,2-dimethylthiazolidine-3-carboxylate (32)

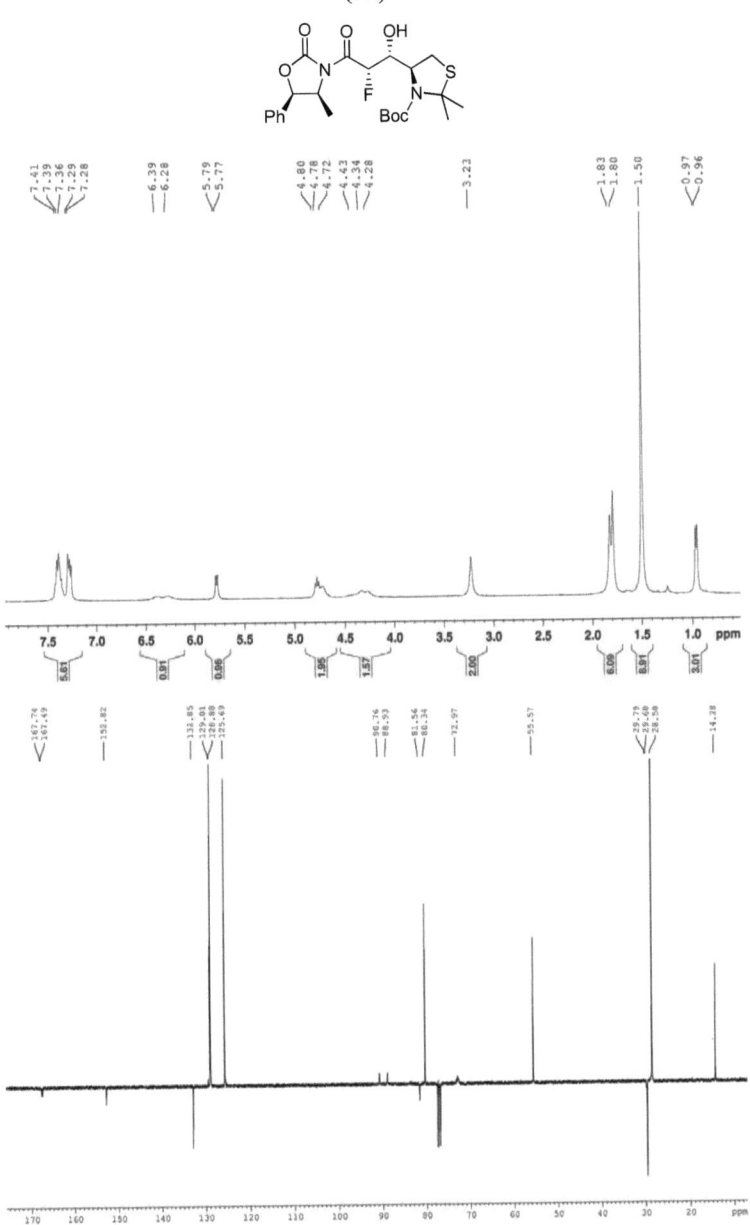

(R)-tert-butyl 4-((1R,2S)-2-fluoro-1-hydroxy-3-((4S,5R)-4-methyl-2-oxo-5-phenyloxazolidin-3-yl)-3-oxopropyl)-2,2-dimethylthiazolidine-3-carboxylate (33)

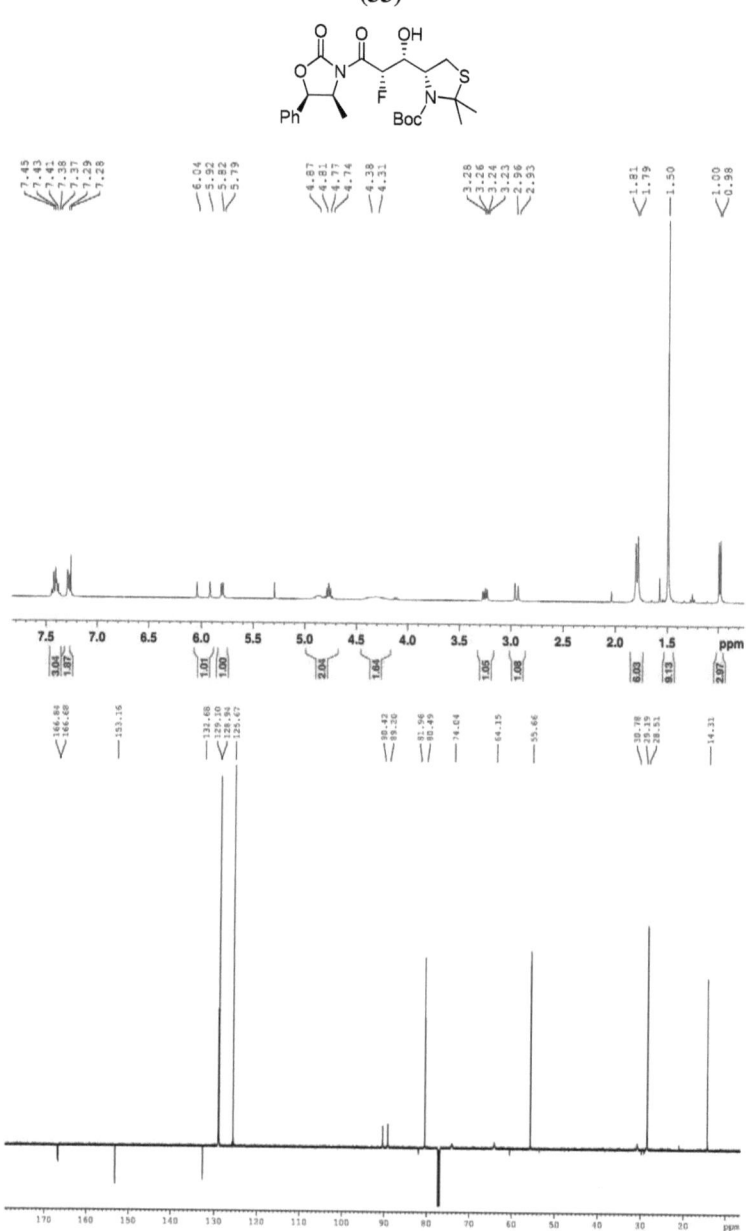

(S)-tert-butyl 4-((1R,2R)-2-fluoro-1,3-dihydroxypropyl)-2,2-dimethylthiazolidine-3-carboxylate (34)

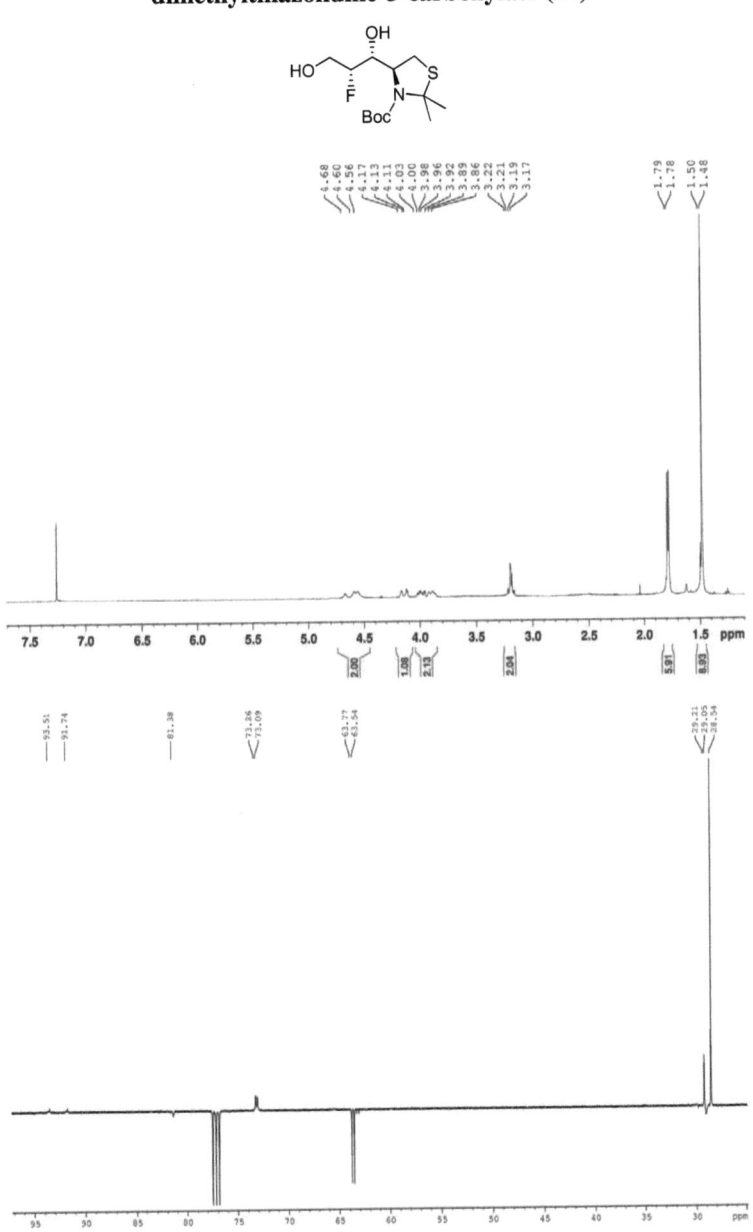

(R)-tert-butyl 4-((1R,2R)-2-fluoro-1,3-dihydroxypropyl)-2,2-dimethylthiazolidine-3-carboxylate (35)

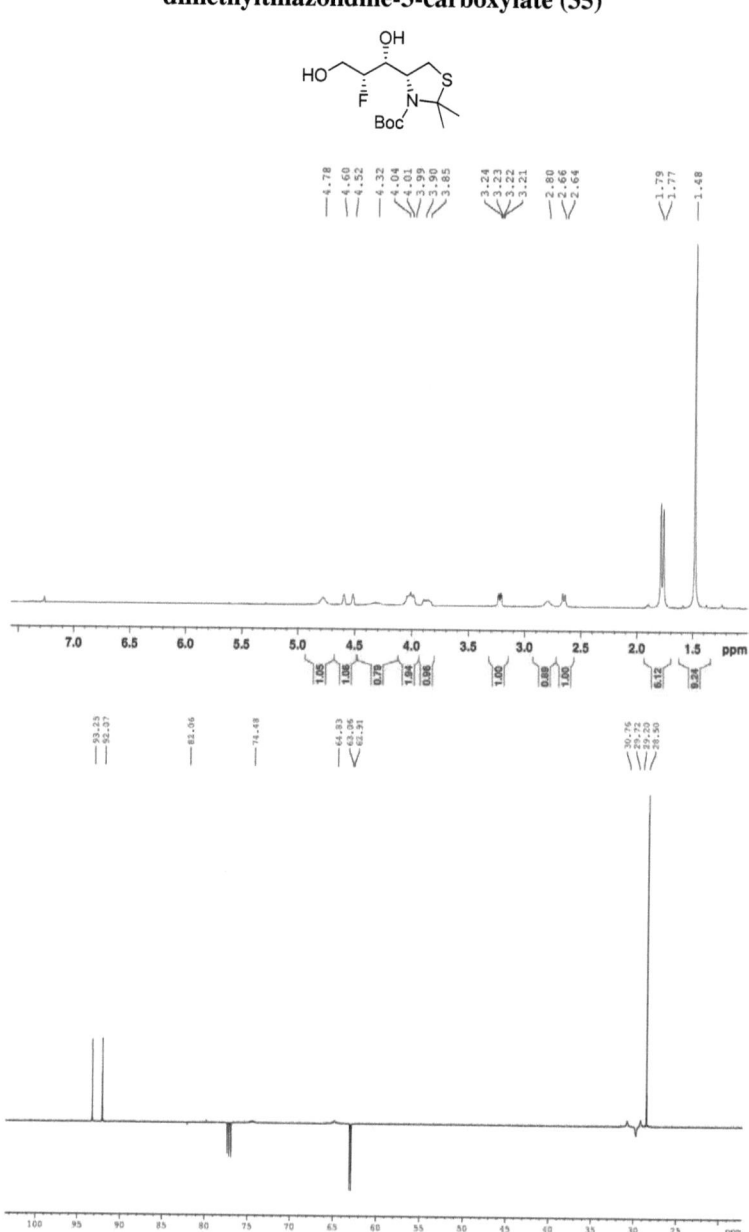

(2R,3R,4R)-4-((2,4-dinitrophenyl)amino)-5-((2,4-dinitrophenyl)thio)-2-fluoropentane-1,3-diol (36)

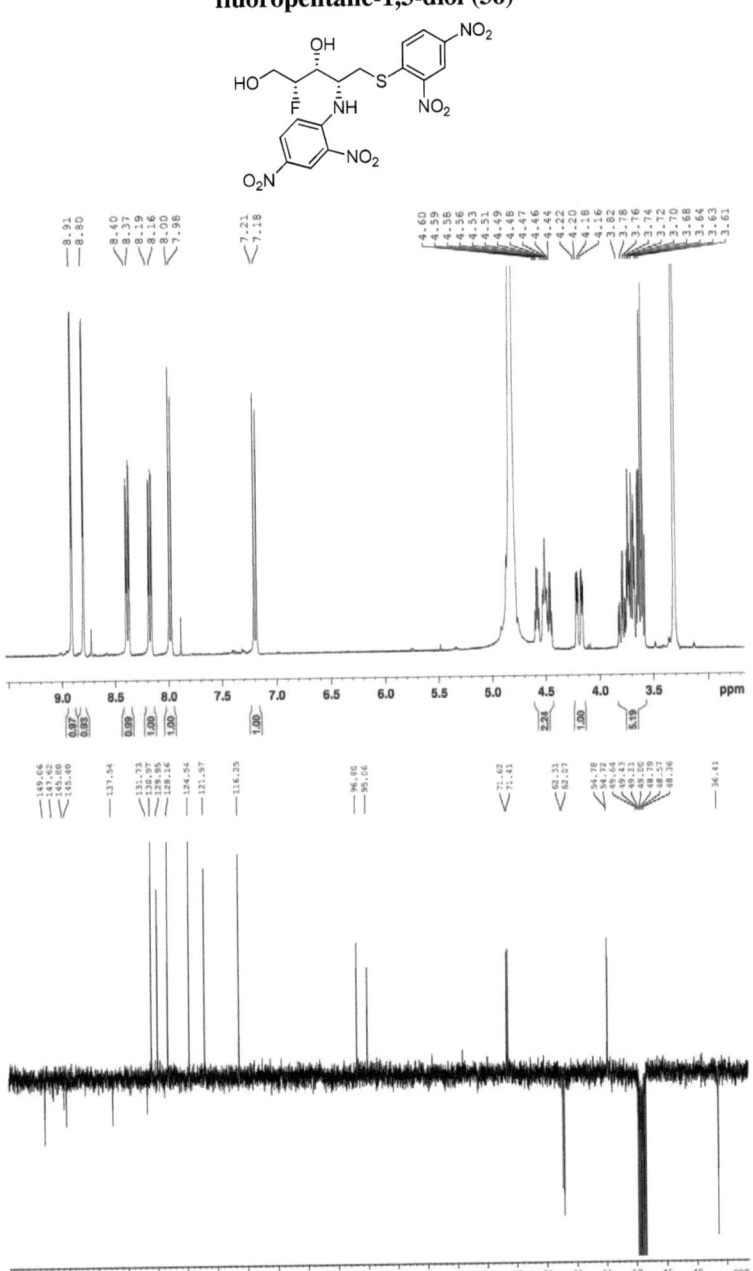

2-acetamido-2,4-dideoxy-4-fluoro-D-lyxose (40)

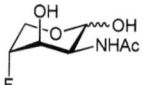

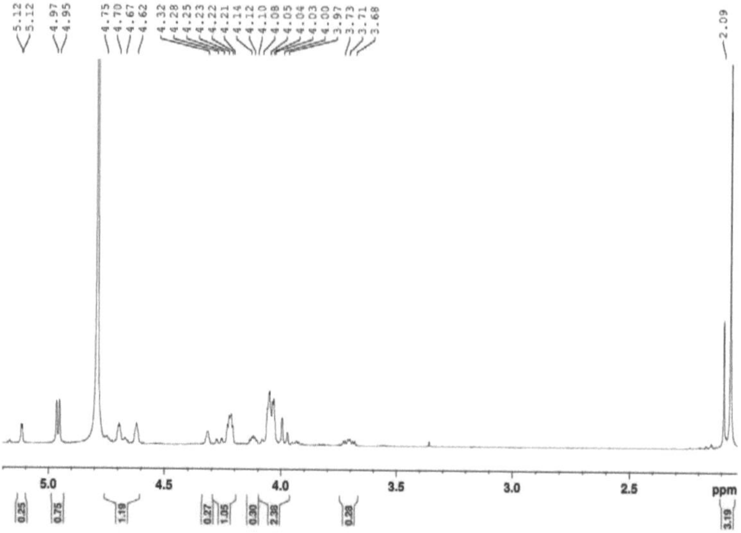

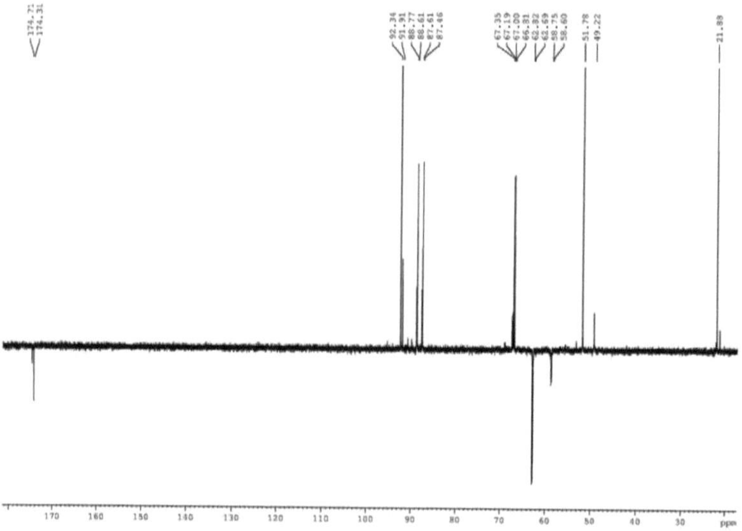

2-acetamido-2,4-dideoxy-4-fluoro-D-xylose (41)

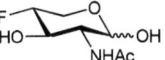

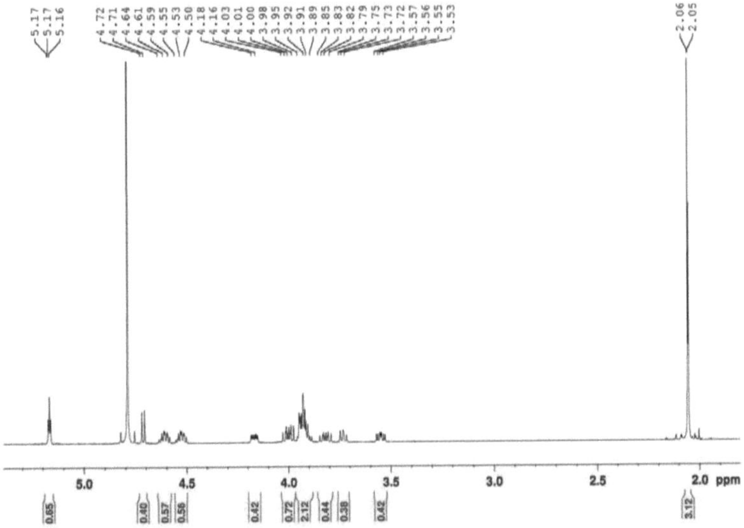

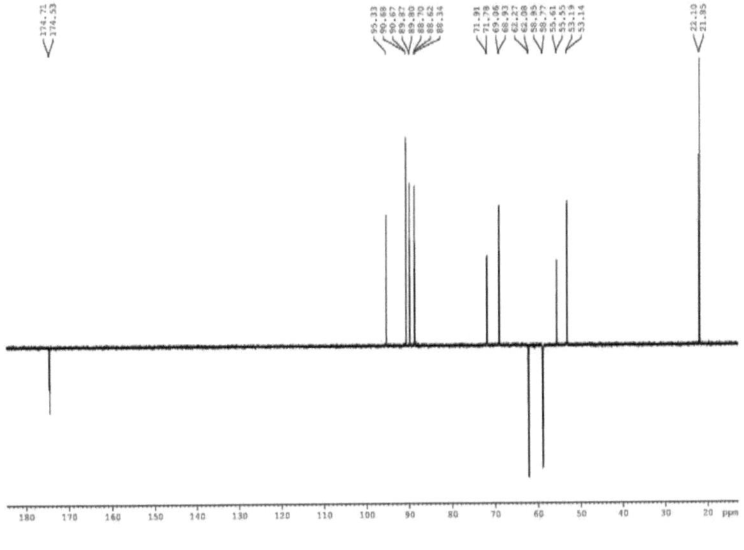

(2R,3R,4R,5R,6S)-6-fluoro-5-hydroxy-7-((4S,5R)-4-methyl-2-oxo-5-phenyloxazolidin-3-yl)-7-oxoheptane-1,2,3,4-tetrayl tetraacetate (42)

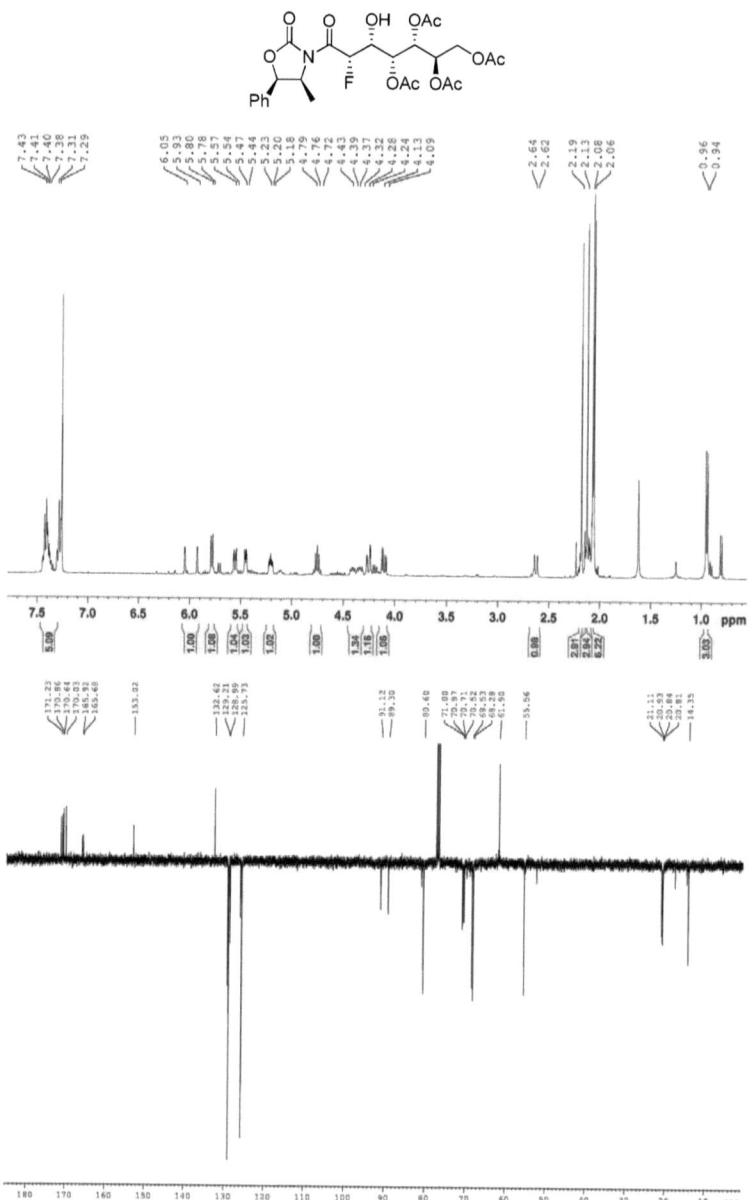

5 References

[1] Albler, C.; Hollaus, R.; Kählig, H.; Schmid, W. *Beilstein J. Org. Chem.* **2014**, 10, 2230-2234.
[2] a) Albler, C.; Schmid, W. *Eur. J. Org. Chem.* **2014**, 2451-2459; b) Albler, C.; Schmid, W. *Eur. J. Org. Chem.* **2015**, 1314-1319. Copyright Wiley–VCH Verlag GmbH & Co. KGaA. Reproduced with permission.
[3] Chao, L.-C.; Rieke, R. D. *J. Organomet. Chem.* **1974**, 67, C64.
[4] Araki, S.; Ito, H.; Butsugan, Y. *J. Org. Chem.* **1988**, 53, 1831.
[5] a) Li, C. J.; Chan, T. H. *Tetrahedron Lett.* **1991**, 32, 7017; b) Li, C.; Chan, T. *Tetrahedron*, **1999**, 55,11149-11176.
[6] Chan, T.H.; Li, C.J.; Lee, M. C.; Wei, Z.Y. *Can. J. Chem.* **1994**, 72, 1181.
[7] Kim, E.; Gordon, D. M.; Schmid, W.; Whitesides, G. M. *J. Org. Chem.* **1993**, 58, 5500.
[8] Chan, T.H.; Yang, Y. *J. Am. Chem. Soc.* **1999**, 121, 3228-3229.
[9] Shen, Z.-L.; Wang, S.Y.; Chok, Y.-K.; Xu, Y.-H.; Loh, T.-P. *Chem. Rev.* **2013**, 113, 271-401.
[10] Augé, J.; Lubin-Germain, N.; Thiaw-Woaye, A. *Tetrahedron Lett.* **1999**, 40, 9245.
[11] Skaanderup, P. R.; Madsen, R. *J. Org. Chem.* **2003**, 68, 2115.
[12] Li, C. J.; Lu, Y. Q. *Tetrahedron Lett.* **1995**, 36, 2721.
[13] Loh, T.-P.; Ho, D. S.-C.; Xu, K.-C.; Sim, K. Y. *Tetrahedron Lett.* **1997**, 38, 865.
[14] a) Paquette, L. A.; Rothhaar, R. R.; Isaac, M.; Rogers, L. M.; Rogers, R.D. *J.Org. Chem.* **1998**, 63, 5463; b) Lobben, P. C.; Paquette, L. A. *J.Org. Chem.* **1998**, 63, 6990; c) Paquette, L. A.; Mendez-Andino, J. L. *J. Org. Chem.* **1998**, 63, 9061.
[15] Zimmerman, H.E.; Traxler, M.D. *J. Am. Chem. Soc.*, **1957**, 79, 1920.
[16] Levoirier, E.; Canac, Y.; Norsikian, S.; Lubineau, A. *Carbohydr. Res.* **2004**, 339, 2737.
[17] a) Loh, T.-P.; Tan, K.-T.; Hu, Q.-Y. *Tet. Lett.* **2001**, 42, 8705; b) Loh, T.-P.; Song, H.-Y. *Synlett* **2002**, 2119.
[18] Paquette, L. A. *Synthesis* **2003**, 765.
[19] Hirashita, T.; Inoue, S.; Yamamura, H.; Kawai, M.; Araki, S. *J.Organomet. Chem.* **1997**, 549, 305.
[20] Araki, S.; Horie, T.; Kato, M.; Hirashita, T.; Yamamura, H.; Kawai, M. *Tet. Lett.* **1999**, 40, 2331.
[21] Lee, P. H.; Anh, H.; Lee, K.; Sung, S.-y.; Kim, S. *Tet. Lett.* **2001**, 42, 37.
[22] Hirayama, L. C.; Gamsey, S.; Knueppel, D.; Steiner, D.; DeLaTorre, K.; Singaram, B. *Tet. Lett.* **2005**, 46, 2315.
[23] Beuchet, P.; Le Marrec, N.; Mosset, P. *Tet. Lett.* **1992**,33, 5959.
[24] Ranu, B. C.; Das, A. *Tet. Lett.* **2004**, 45, 6875.
[25] Yadav, J. S.; Anjaneyulu, S.; Ahmed, M. M.; Reddy, B. V. S. *Tet. Lett.* **2001**, 42, 2557.
[26] Warwel, M.; Fessner, W.-D. *Synlett* **2000**, 865.
[27] Chan, T. H.; Lee, M. C. *J. Org. Chem.* **1995**, 60, 4228.
[28] a) Chan, T.-H.; Xin, Y.-C *Chem. Commun.* **1996**, 905; b) Chan, T.-H.; Xin, Y.-C.; von Itzstein, M. *J. Org. Chem.* **1997**, 62, 3500.
[29] Gao, J.; Harter, R.; Gordon, D. M.; Whitesides, G. M. *J. Org. Chem.* **1994**, 59, 3714.
[30] Lee, Y. J.; Kubota, A.; Ishiwata, A.; Ito, Y. *Tet. Lett.* **2011**, 52, 418–421.
[31] Thibault, P.; Logan, S. M.; Kelly, J. F.; Brisson, J.-R.; Ewing, C. P.; Trust, T. J.; Guerry, P. *J. Biol. Chem.* **2001**, 276, 34862–34870.
[32] Schmölzer, C.; Nowikow, C.; Kählig, H.; Schmid, W. *Carb. Res.* **2013**, 367, 1-4.
[33] Kim, E.; Gordon, D. M.; Schmid, W.; Whitesides, G. M. *J. Org. Chem.* **1993**, 58, 5500.
[34] Miao, W.; Chung, L. W.; Wu, Y.-D.; Chan, T. H. *J. Am. Chem. Soc.* **2004**, 126, 13326.
[35] Isaac, M. B.; Chan, T.-H. *J. Chem. Soc., Chem. Commun.* **1995**, 1003.
[36] Lin, M. J.; Loh, T. P. *J. Am. Chem. Soc.* **2003**, 125, 13042.
[37] Fischer, M.; Schmölzer, C.; Nowikow, C.; Schmid, W. *Eur. J. Org. Chem.* **2011**, 1645.
[38] a) Lee, P. H.; Kim, H.; Lee, K.; Kim, M.; Noh, K.; Kim, H.; Seomoon, D. *Angew. Chem., Int. Ed.* **2005**, 44, 1840; b) Park, J.; Lee, P. H. *Org. Lett.* **2008**, 10, 3359; c) Samanta, K.; Kar, G. K.; Sarkar, A. K. *Tetrahedron Lett.* **2012**, 53, 1376.
[39] Waterworth, L.; Worrall, I. J. *Chem. Commun.* **1971**, 569.
[40] Shen, Z. L.; Goh, K. K. K.; Yang, Y. S.; Lai, Y. C.; Wong, C. H. A.; Cheong, H. L.; Loh, T. P. *Angew. Chem., Int. Ed.* **2011**, 50, 511.

[41] Wittig, G., Frommeld, H. D.; Suchanek, P. *Angew. Chem.* **1963**, 75, 978–979.
[42] Mukaiyama, T.; Narasaka, K.; Banno, K. *Chem. Lett.* **1973**, 1011–1014.
[43] House, H. O.; Crumrine, D. S.; Teranishi, A. Y.; Olmstead, H. D. *J. Am. Chem. Soc.* **1973**, 95(10), 3310–24.
[44] Saigo, K.; Osaki, M.; Mukaiyama, T. *Chem. Lett.* **1975**, 989–990.
[45] Mukaiyama, T.; Ishida, A. *Chem. Lett.* **1975**, 319–322.
[46] Kalesse, M.; Cordes, M.; Symkenberga, G.; Lu, H.-H. *Nat. Prod. Rep.* **2014**, 31, 563–594.
[47] Murata, S.; Suzuki, M.; Noyori, R. *J. Am. Chem. Soc.* **1980**, 102, 3248–3249.
[48] a) Masamune, S.; Sato, T.; Kim, B. M.; Wollmann, T. A. *J. Am. Chem. Soc.* **1986**, 108, 8279–8281; b) Paterson, I.; Lister, M. A.; McClure, C. K. *Tetrahedron Lett.* **1986**, 27, 4787–4790; c) Corey, E. J.; Imwinkelried, R.; Pikul, S.; Xiang, Y. B. *J. Am. Chem. Soc.* **1989**, 111, 5493–5495.
[49] a) Iwasawa, N.; Mukaiyama, T. *Chem. Lett.* **1982**, 1441–1444; b) Iwasawa, N.; Mukaiyama, T. *Chem. Lett.* **1983**, 297–298; c) Mukaiyama, T.; Iwasawa, N.; Stevens, R. W.; Haga, T. *Tetrahedron* **1984**, 40, 1381-1390.
[50] a) Boxer, M. B.; Yamamoto, H. *J. Am. Chem. Soc.* **2007**, 129, 2762–2763; b) Boxer, M. B.; Akakura, M.; Yamamoto, H. *J. Am. Chem. Soc.* **2008**, 130, 1580–1582; c) Saadi, J.; Akakura, M.; Yamamoto, H. *J. Am. Chem. Soc.* **2011**, 133, 14248–14251.
[51] Shirokawa, S.-i.; Kamiyama, M.; Nakamura, T.; Okada, M.; Nakazaki, A.; Hosokawa, S.; Kobayashi, S. *J. Am. Chem. Soc.* **2004**, 126, 13604–13605.
[52] a) Nicolaou, K. C.; Guduru, R.; Sun, Y.-P.; Banerji, B.; Chen, D. K. *Angew. Chem., Int. Ed.* **2007**, 46, 5896-5900; b) Nicolaou, K. C.; Sun, Y.-P.; Guduru, R.; Banerji, B.; Chen, D. K. *J. Am. Chem. Soc.* **2008**, 130, 3633–3644; c) Jiang, X.; Liu, B.; Lebreton, S.; De Brabander, J. K. *J. Am. Chem. Soc.* **2007**, 129, 6386-6387.
[53] Shinoyama, M.; Shirokawa, S.-i.; Nakazaki, A.; Kobayashi, S. *Org. Lett.* **2009**, 11, 1277–1280.
[54] Wang, L.; Gong, J.; Deng, L.; Xiang, Z.; Chen, Z.; Wang, Y.; Chen, J.; Yang, Z. *Org. Lett.* **2009**, 11, 1809-1812
[55] Wang, L.; Xi, Y.; Yang, S.; Zhu, R.; Liang, Y.; Chen, J.; Yang, Z. *Org. Lett.* **2011**, 13, 74–77.
[56] Mukaeda, Y.; Kato, T.; Hosokawa, S. *Org. Lett.* **2012**, 14, 5298–5301.
[57] Symkenberg, G.; Kalesse, M. *Org. Lett.* **2012**, 14, 1608–1611.
[58] Lisboa, M. P.; Jones, D. M.; Dudley, G. B. *Org. Lett.* **2013**, 15, 886–889.
[59] Evans, D. A.; Nelson, J. V.; Vogel, E.; Taber, T. R. *J. Am. Chem. Soc.* **1981**, 103, 3099-3111.
[60] Heathcock, C. H.; Buse, C. T.; Kleschick, W. A.; Pirrung, M. C.; Sohn, J. E.; Lampe, J. *J. Org. Chem.* **1980**,45, 1066-1081.
[61] Evans, D. A.; Downey, C. W.; Shaw, J. T.; Tedrow, J. S. *Org. Lett.* **2002**, 4, 1127-1130.
[62] a) Li, Y.; Paddon-Row, M. N.; Houk, K. N. *J. Org. Chem.* **1990**, 55, 481–493; b) Denmark, S. E.; Henke, B. R. *J. Am. Chem. Soc.* **1991**, 113, 2177–2194; c) Geary, L. M.; Hultin, P. G. *Tetrahedron: Asymmetry* **2009**, 20, 131–173.
[63] Evans, D. A.; Bartroli, J.; Shih, T. L. *J. Am. Chem. Soc.* **1981**, 103, 2127-2129.
[64] a) Cowden, C. J.; Paterson, I.; Paquette, L. A. Ed. *Organic Reactions*, Wiley, **1997**, Vol. 51 (Chapter 1); b) Franklin, A. S.; Paterson, I. *Contemp. Org. Synth.* **1994**, 1, 317–338; c) Ager, D. J.; Prakash, I.; Schaad, D. R. *Aldrichim. Acta* **1997**, 30, 3–12.
[65] a) Evans, D. A.; Bender, S. L.; Morris, J. *J. Am. Chem. Soc.* **1988**, 110, 2506-2526; b) Heravi, M. M.; Zadsirjan, V. *Tetrahedron: Asymmetry* **2013**, 1149–1188.
[66] Zappia, G.; Gacs-Baitz, E. G.; Monache, D.; Misiti, D.; Nevola, L.; Botta, B. *Curr. Org. Synth.* **2007**, 4, 81-135.
[67] a) Nagao, Y.; Hagiwara, Y.; Kumagai, T.; Ochiai, M.; Inoue, T.; Hashimoto, K.; Fujita, E. *J. Org. Chem.* **1986**, 51, 2391-2393; b) Hsiao, C.-N.; Liu, L.; Miller, M. J. *J. Org. Chem.* **1987**, 52, 2201-2206.
[68] Evans, D. A.; Rieger, D. L.; Bilodeau, M. T.; Urpi, F. *J. Am. Chem. Soc.* **1991**, 113, 1047-1049.
[69] a) Yan, T.-H.; Tan, C.-W.; Lee, H.-C.; Lo, H.-C.; Huang, T.-Y. *J. Am. Chem. Soc.* **1993**, 115, 2613-2621; b) Yan, T.-H.; Hung, A.-W.; Lee, H.-C.; Chang, C.-S.; Liu, W.-H. *J. Org. Chem.* **1995**, 60, 3301-3306.
[70] Crimmins, M. T.; King, B. W.; Tabet, E. A. *J. Am. Chem. Soc.* **1997**, 119, 7883-7884.
[71] Walker, M. A.; Heathcock, C. H. *J. Org. Chem.* **1991**, 56, 5747-5750.

[72] Pridgen, L. N.; Abdel-Magid, A. F.; Lantos, I.; Shilcrat, S.; Eggleston, D. S. *J. Org. Chem.* **1993**, 58, 5107-5117.
[73] a) Abdel-Magid, A.; Lantos, I.; Pridgen, L. N. *Tetrahedron Lett.* **1984**, 25, 3273; b) Abdel-Magid, A.; Pridgen, L. N.; Eggleston, D. S.; Lantos, I. *J. Am. Chem. Soc.* **1986**, 108, 4595.
[74] Garner, P. *Tetrahedron Lett.* **1984**, 25, 5855.
[75] Garner, P.; Park, J. M. *J. Org. Chem.* **1987**, 52, 2361-2364.
[76] McKillop, A.; Taylor, R. J. K.; Watson, R. J.; Lewis, N. *Synthesis* **1994**, 31.
[77] Roush, W. R.; Hunt, J. A.; *J. Org Chem.* **1995**, 60, 798.
[78] Dondoni, A.; Perrone, D. *Synthesis* **1997**, 527.
[79] a) Casiraghi, G.; Colombo, L.; Rassu, G.; Spanu, P. *J. Chem. Soc.,Chem. Commun.* **1991**, 603; b) Casiraghi, G.; Colombo, L.; Rassu, G.; Spanu, P. *J. Org. Chem.* **1991**, 56, 6523.
[80] a) Marcus, J.; Vandermenten, G. W. M.; Brussee, J.; van derGren, A. *Tetrahedron:Asymmetry* **1999**, 10, 1617; b) Soro, P.; Rassu, G.; Spanu, P.; Pinna, L.; Zanardi, F.; Casiraghi, G. *J. Org.Chem.* **1996**, 61, 5172.
[81] Herold, P. *Helv. Chim. Acta*, **1988**, 71, 354.
[82] Liang, X.; Andersch, J.; Bols, M. *J. Chem. Soc., Perkin Trans. 1* **2001**, 2136–2157.
[83] a) Mohapatra, D. K.; Mondal, D.; Chorghadeb, M. S.; Gurjara, M. K. *Tet. Lett.* **2006**, 47, 9215-9219; b) Mohapatra, D. K.; Mondal, D.; Gonnade, R. G.; Chorghade, M. S.; Gurjar, M. K. *Tet. Lett.* **2006**, 47, 6031-6035.
[84] a) Loew, O. *Dtsch. Chem. Ges.* **1889**, 22, 478; b) Fischer, E. Passmore, F. *Dtsch. Chem. Ges.* **1890**, 23, 370.
[85] Müller, D.; Pitsch, S.; Kittaka, A.; Wagner, E.; Wintner, C.E.; Eschenmoser, A. *Helv. Chim. Acta* **1990**, 73, 1410.
[86] Weber, A. L. *J. Mol. Evol.* **1992**, 35, 1.
[87] Pitsch, S.; Krishnamurthy, R.; Arrhenius, G. *Helv. Chim. Acta* **2000**, 83, 2398.
[88] Guanti, G.; Banfi, L.; Zannetti, M. T. *Tet. Lett.* **2000**, 41, 3181.
[89] Vogel, P.; Robina, I.; Fraser-Reid, B.; Tatsuta, K. (eds) *GlycoscienceChemistry and Chemical Biology*, Springer, Berlin Heidelberg New York 2008, 20.
[90] Suri, J. T.; Ramachary, D. B.; Barbas, C. F. *Org. Lett.* **2005**, 7, 1383.
[91] Enders, D. Grondal, C. *Angew. Chem. Int. Ed.* **2005**, 44, 1210.
[92] List, B. *J. Am. Chem. Soc.* **2000**, 122, 9336.
[93] Enders, D.; Grondal, C.; Vrettou, M.; Raabe, G. *Angew. Chem. Int. Ed.* **2005**, 44, 4079.
[94] Ibrahem, I.; Zou, W. B.; Casas, J.; Sunden, H.; Córdova, A. *Tetrahedron* **2006**, 62, 357.
[95] Hayashi, Y.; Tsuboi, W.; Ashimine, I.; Urushima, T.; Shoji, M.; Sakai, K. *Angew. Chem. Int. Ed.* **2003**, 42, 3677.
[96] Mukaiyama, T.; Shiina, I.; Kobayashi, S. *Chem. Lett.* **1990**, 2201.
[97] Kobayashi, S.; Kawasuji, T. *Synlett* **1993**, 911.
[98] Braun, M.; Moritz, J. *Synlett* **1991**, 750.
[99] Evans, D. A.; Gage, J. R.; Leighton, J. L. *J. Am. Chem. Soc.* **1992**, 9434.
[100] a) Prenner, R. H.; Binder, W. H.; Schmid, W. *Liebigs Ann. Chem.* **1994**, 73; b) Binder, W. H.; Prenner, R. H.; Schmid, W. *Tetrahedron* **1994**, 50, 749; c) Schmölzer C.; Fischer M.; Schmid W. *Eur. J. Org. Chem.* **2010**, 4886.
[101] a) Hollaus, R. *Indium unterstützte Allylierung in der Synthese Stickstoff-haltiger höherer Kohlenhydrate.* Master Thesis, University of Vienna, AUT, **2009**; b) Albler, C. *Synthese von höheren C-Glycosiden.* Master Thesis, University of Vienna, AUT, **2011**.
[102] Cresswell, A. J.; Davies, S. G.; Lee, J. A.; Morris, M. J.; Roberts, P. M.; Thomson, J. E. *J. Org. Chem.* **2012**, 7262.
[103] Marigo, M.; Fielenbach, D.; Braunton, A.; Kjœrsgaard, A.; Jørgenson, K. A. *Angew. Chem.* **2005**, 117, 3769-3772.
[104] Beeson, T. D.; MacMillan, D. W. C. *J. Am. Chem. Soc.* **2005**, 127, 8826-8828.
[105] a) Roush, W. R.; Hunt, J. A. *J. Org. Chem.* **1995**, 60, 798-806; b) Lemke, A.; Büschleb, M.; Ducho, C. *Tetrahedron* **2010**, 66, 208-214.

[106] Matt, T.; Leong, C.; Lang, K.; Sha, S.; Akbergenov, R.; Shcherbakov, D.; Meyer, M.; Duscha, S.; Xie, J.; Dubbaka, S. R.; Perez-Fernandez, D.; Vasella, A.; Ramakrishnan, V.; Schacht, J.; Böttger, E. C. *Proc. Natl. Acad. Sci. U. S. A.* **2012**, 10984-10989.

[107] a) Amuras, J.-I.; Horito,S.; Hashimoto, H.; Yoshimura, J. *Carb. Res.* **1988**, 181-199; b) J. A. Marshall, S. Beaudoin, *J. Org. Chem.* 1996, 61, 581.

[108] Fair, R. J.; Hensler, M. E.; Thienphrapa, W.; Dam, Q. N.; Nizet, V.; Tor, Y. *ChemMedChem* **2012**, 1237-1244.

[109] Kuhn, R.; Kirschenlohr, W. *Angew. Chem.* **1955**, 786.

[110] a) Perez, J. A. G.; Corraliza, R. M. P.; Galan, E. R.; Guillen, M. G. *Ann. Quim.* **1979**, 387-391; b) Albarran, J. C. P.; Galan, E. R.; Perez, J. A. G. *Carb. Res.* **1985**, 117-128.

[111] Martin, O. R.; Szarek, W. A. *Carb. Res.* **1984**, 195-219.

[112] Simonsen, K. B.; Ayida, B. K.; Vourloumis, D.; Takahashi, M.; Winters, G. C.; Barluenga, S.; Qamar, S.; Shandrick, S.; Zhao, Q.; Hermann, T. *ChemBioChem* **2002**, 1223-1228.

[113] a) Seltmann, G.; Holst, O. *The Bacterial Cell Wall*, Springer, Berlin, **2002**; b) Moran, A. P.; Holst, O.; Brennan, P. J.; von Itzstein, M. *Microbial Glycobiology*, Academic Press, London, **2009**.

[114] Wanty, C.; Anandan, A.; Piek, S.; Walshe, J.; Ganguly, J.; Carlson, R. W.; Stubbs, K. A.; Kahler, C. M.; Vrielink, A. *J. Mol. Biol.* **2013**, 425, 3389–3402.

[115] Liao, X.; Poirot, E.; Chang, A. H. C.; Zhang, X.; Zhang, J.; Nato, F.; Fournier, J.-M.; Kováč, P.; Glaudemans, C. P. J. *Carbohydr. Res.* **2002**, 337, 2437–2442.

[116] a) Danac, R.; Ball, L.; Gurr, S. J.; Muller, T.; Fairbanks A. J. *ChemBioChem* **2007**, 8, 1241 – 1245; b) Nigro, J.; Wang, A.; Mukhopadhyay, D.; Lauer, M.; Midura, R. J.; Sackstein, R.; Hascall, V. C. *J. Biol. Chem.* **2009**, 284, 16832-16839; c) Xue, J.; Kumar, V.; Khaja, S. D.; Chandrasekaran, E. V.; Locke, R. D.; Matta K. L. *Tetrahedron* **2009**, 65, 8325–8335; d) Li, Y.; Zhou, Y.; Mac, Y.; Li, X. *Carb. Res.* **2011**, 346, 1714-1720.

[117] O'Hagan, D. *Chem. Soc. Rev.* **2008**, 37, 308–319.

[118] a) André, S.; Cañada, F. J.; Shiao, T. C.; Largartera, L.; Diercks, T.; Bergeron-Brlek, M.; Biari, K.; Papadopoulos, A.; Ribeiro, J. P.; Touaibia, M.; Solís, D.; Menéndez, M.; Jiménez-Barbero, J.; Roy, R.; Gabius, H.-J. *Eur. J. Org. Chem.* **2012**, 4354–4364; b) Braitsch, M.; Kählig, H.; Kontaxis, G.; Fischer, M.; Kawada, T.; Konrat, R.; Schmid, W. *Beilstein J. Org. Chem.* **2012**, *8*, 448–455; c) Aspers, R. L. E. G.; Ampt, K. A. M.; Dvortsak, P.; Jaeger, M.; Wijmenga, S. S. *J. Magn. Reson.* **2013**, *231*, 79–89.

[119] Chang, A. H. C.; Horton, D.; Kováč, P. *Tetrahedron: Asymmetry*, **2000**, 11, 595-606.

[120] a) Marigo, M.; Franze´n, J.; Poulsen, T. B.; Zhuang, W.; Jørgensen, K. A. *J. Am. Chem. Soc.* **2005**, 127, 6964; b) Zhao, G.-L.; Ibrahim, I.; Sundén, H.; Córdova A. *Adv. Synth. Catal.* **2007**, 349, 1210.

[121] Miyashita, M.; Mizutani, T.; Tadano, G.; Iwata, Y.; Miyazawa, M.; Tanino, K. *Angew. Chem. Int. Ed.* **2005**, 44, 5094.

[122] a) Trost, B. M.; Bunt, R. C.; Lemoine, R. C.; Calkins, T. L. *J. Am. Chem. Soc.* **2000**, 5968-5976; b) Trost, B. M.; Jiang, C.; Hammer, K. *Synthesis* **2005**, 3335-3345; c) Trost, B. M.; Horne, D. B.; Woltering, M. J. *Chem. Eur. J.* **2006**, 6607-6620.

[123] Etter, M. C.; Baures, P. W. *J. Am. Chem. Soc.* **1988**, 110, 639-640.

[124] a) Bhacca, N. S.; Horton, D.; Paulsen, H. *J. Org. Chem.* **1968**, 2484-2487; b) Horita, D. A.; Hajduk, P. J.; Lerner, L. E. *Glycoconjugate J.* **1997**, 691-696; c) Kräutler, V.; Müller, M.; Hünenberger, P. H. *Carb. Res.* **2007**, 2097-2124.

[125] Karplus, M. *J. Am. Chem. Soc.* **1963**, 85 (18), 2870-2871.

[126] Rastogi, S. K.; Kornienko, A. *Tetrahedron: Asymmetry*, **2006**, 17, 3170-3178.

[127] Katsuki, T.; Sharpless, K. B. *J. Am. Chem. Soc.* **1980**, 102 (18), 5974-5976.

[128] Reetz, M. T.; Lauterbach, E. H. *Tet. Lett.* **1991**, 32 (35), 4477-4480.

[129] Roush, W. A.; Walts, A. E.; Hoong, L. K. *J. Am. Chem. Soc.* **1985**, 107, 8186-8190.

[130] Tamborini, L.; Conti, P.; Pinto, A.; Colleoni, S.; Gobbi, M.; De Micheli, C. *Tetrahedron* **2009**, 65 (31),6083-6089.

[131] Hulme, A. N.; Montgomery, C. H.; Henderson, D. K. *J. Chem. Soc., Perkin Trans. 1*, **2000**, 1837-1841.

[132] Ayad, T.; Génisson; Y.; Broussy, S.; Baltas, M.; Gorrichon, L. *Eur. J. Org. Chem.* **2003**, 2903-2910.
[133] Duthaler, R. O.; Wyss, B. *Eur. J. Org. Chem.* **2011**, 4667–4680.
[134] a) Takata, T.; Tamura, Y.; Ando, W. *Tetrahedron* **1985**, 11, 2133-2137; b) Lu, X.; Long, T. E. *Tet. Lett.* **2011**, 52, 5051–5054.
[135] Iwakawa, M.; Pinto, B. M.; Szarek, W. A. *Can. J. Chem.* **1978**, 56, 326–335.
[136] Tokitoh, N.; Igarashi, Y.; Ando, W. *Tet. Lett.* **1987**, 28, 5903–5906.
[137] Takata, T.; Tamura, Y.; Ando, W. *Tetrahedron* **1985**, 41, 2133–2137.
[138] Turro, N. J.; Ramamurthy, V.; Scaiano, J. *Modern Molecular Photochemistry of Organic Molecules*, Palgrave Macmillan, **2010**.
[139] Craig, N. C.; Chen, A.; Suh, K. H.; Klee, S.; Mellau, G. C.; Winnewisser, B. P.; Winnewisser, M. *J. Am. Chem. Soc.* **1997**, 119 (20), 4789 – 4790.
[140] Dohi, H.; Périon, R.; Durka, M.; Bosco, M.; Roué, Y.; Moreau, F.; Grizot, S.; Ducruix, A.; Escaich, S.; Vincent, S. P. *Chem. Eur. J.* **2008**, 14, 9530-9539.
[141] Wolfrom, M. L.; Weisblat, D. I.; Zophy, W. H.; Waisbrot, S. W. *J. Am. Chem. Soc.* **1941**, 63, 201-203.

I want morebooks!

Buy your books fast and straightforward online - at one of the world's fastest growing online book stores! Environmentally sound due to Print-on-Demand technologies.

Buy your books online at
www.get-morebooks.com

Kaufen Sie Ihre Bücher schnell und unkompliziert online – auf einer der am schnellsten wachsenden Buchhandelsplattformen weltweit! Dank Print-On-Demand umwelt- und ressourcenschonend produziert.

Bücher schneller online kaufen
www.morebooks.de

OmniScriptum Marketing DEU GmbH
Heinrich-Böcking-Str. 6-8
D - 66121 Saarbrücken
Telefax: +49 681 93 81 567-9

info@omniscriptum.com
www.omniscriptum.com

Printed by Books on Demand GmbH, Norderstedt / Germany